AF514246

Matthias Dominik Wissert

Optical Antennas

Linear and Nonlinear Excitation and Emission

Optical Antennas

Linear and Nonlinear Excitation and Emission

by
Matthias Dominik Wissert

KIT Scientific Publishing

Dissertation, Karlsruher Institut für Technologie
Fakultät für Elektrotechnik und Informationstechnik, 2011

Impressum

Karlsruher Institut für Technologie (KIT)
KIT Scientific Publishing
Straße am Forum 2
D-76131 Karlsruhe
www.ksp.kit.edu

KIT – Universität des Landes Baden-Württemberg und nationales
Forschungszentrum in der Helmholtz-Gemeinschaft

KIT Scientific Publishing 2012
Print on Demand

ISBN 978-3-86644-765-3

Optical Antennas: Linear and Nonlinear Excitation and Emission

Zur Erlangung des akademischen Grades eines DOKTOR-INGENIEURS

von der Fakultät für

Elektrotechnik und Informationstechnik
des Karlsruher Instituts für Technologie (KIT)

genehmigte

DISSERTATION

von

Dipl.-Ing. Matthias Dominik Wissert

geb. in Stuttgart

Tag der mündlichen Prüfung: 12. Juli 2011
Hauptreferent: Prof. Dr. rer. nat. Uli Lemmer
Korreferent: Prof. Dr. rer. nat. Michael Siegel

Zusammenfassung

Optische Antennen sind metallische Nanostrukturen, die bei optischen Wellenlängen Resonanzverhalten zeigen. Ähnlich zu makroskopischen Antennen, wie sie beispielsweise in der Radiowellenübertragung Verwendung finden, ist ihr Sende- und Empfangsverhalten von der Größe der Struktur abhängig. Bei genaueren Untersuchungen zeigen sich jedoch deutliche Abweichungen vom Radioantennenverhalten und zusätzliche Einflüsse. Ein Hauptgrund dafür ist, dass Metalle bei optischen Frequenzen keine idealen Leiter mehr sind, so dass von einer vollständigen Verdrängung des elektromagnetischen Feldes im Antenneninneren nicht mehr ausgegangen werden kann.

In dieser Arbeit werden optische Antennen aus Gold und Aluminium auf ihr Resonanzverhalten untersucht. Insbesondere ist von Interesse, wie sich die Kopplung zweier Antennenarme durch einen kleinen Spalt auf das Abstrahlverhalten auswirkt und wie die beiden gewählten Materialien die Antenneneigenschaften beeinflussen.

Der entwickelte Herstellungsprozess für optische Antennen basierend auf Elektronenstrahllithographie und dem anschließenden Aufdampfen von Gold beziehungsweise Aluminium wird vorgestellt. Die entstandenen Nanostrukturen mit minimaler Ausdehnung von etwa 20 nm werden mittels Elektronenmikroskopie und Rasterkraftmikroskopie ausführlich auf ihre topographischen Eigenschaften untersucht.

Die vielfältigen Charakterisierungsmöglichkeiten durch den ebenfalls entwickelten Messaufbau werden im Anschluss dargestellt. Dabei liegt ein besonderes Augenmerk auf der Möglichkeit, Dunkelfeldspektroskopie sowie Konfokalmikroskopie anzuwenden. Die Leistungsfähigkeit des Messplatzes für die Charakterisierung wird durch das Vermessen der Punktspreizfunktion des Systems an einem Quantenpunkt bei Zweiphotonenanregung sowie Messungen zum Einzelphotonenemissionsverhalten von Quantenpunkten (Antibunching) bei solcher Anregung demonstriert.

Mittels Dunkelfeldspektroskopie werden lineare Streuspektren optischer Antennen aus Gold und Aluminium vermessen. Dabei zeigt sich, dass beide Materialien für optische Antennen verwendet werden können. Die Resonanzwellenlänge von Aluminium- und Goldantennen skaliert linear mit der Antennenlänge. Zweiarmantennen mit Spalt zeigen weiterhin eine spektrale Rotverschiebung im Vergleich zu Einarmantennen. Für Goldantennen wird experimentell gezeigt, dass Zweiarmantennen mit Antennenspalt in ihren Abstrahleigenschaften Einarmantennen überlegen sind. Mit optischen Antennen aus Gold werden kleinste Resonanzwellenlängen von etwa 580 nm erreicht. Aluminiumantennen können für kleinere Wellenlängen als Goldantennen eingesetzt werden, allerdings sind sie für Wellenlängen über 700 nm durch Interbandübergänge begrenzt. Im Wellenlängebereich, in dem beide Materialien eingesetzt werden können, wird für Aluminium die größere Antennenarmlänge benötigt. Weiterhin zeigen Aluminiumantennen eine größere spektrale Breite und ihre Güte als Resonator ist geringer.

Gold- sowie Aluminiumantennen werden auch mit gepulstem Laserlicht der Wellenlänge 810 nm angeregt, was zu nichtlinearen Effekten führt. Für günstig gewählte Anregungsintensitäten zeigen Goldnanoantennen (Strukturen sowohl mit als auch ohne Antennenspalt) durch Zweiphotonenabsorption und nachfolgende Plasmonemission Abstrahleigenschaften sehr ähnlich denen bei linearer Streuanregung, obwohl sie nur bei einer Wellenlänge angeregt werden. Diese Zweiphotonenlumineszenz wird durch angeregte elektronische Zustände im Metall, die strahlend unter Erzeugung eines Partikelplasmons relaxieren, erklärt. Für Aluminium werden ähnliche Effekte beobachtet. Neben der physikalischen Bedeutung hat das Verfahren auch praktische Anwendungen. So können mit dieser Charakterisierungsmethode Antennen schnell auf ihre Resonanzeigenschaften bei einer bestimmten Wellenlänge untersucht werden, da die abgestrahlte Intensität mit der Anpassung der Antenne auf die eingestrahlte Lichtwellenlänge skaliert.

Mittels einer neu entwickelten aperturlosen Nahfeldmikroskopiemethode wird außerdem das Nahfeld von Zweiarmantennen mit Spalt unter Zweiphotonenanregung vermessen. Es wird nachgewiesen, dass diese eine erhöhte Nahfeldintensität im Antennenspalt sowie in abgeschwächter Form auch an den außen liegenden jeweiligen Stabenden zeigen. Die Untersuchung von nichtresonanten Einarmantennen, ebenfalls unter Zweiphotonenanregung, führt durch sogenannte Blitzableitereffekte auf eine räumlich deutlich unterhalb der Beugungsgrenze liegende Lichtantwort. Diese Konzentration elektromagnetischer Felder wird zusätzlich mittels Quantenpunkten, die als lokale Sonden eingesetzt werden, nachgewiesen. Durch die hohe Felddichte kommt es außerdem zu Verformungen der Strukturen, welche mittels Rasterkraftmikroskopie analysiert werden.

Zusammengefasst führen die beobachteten Effekte zu einem größeren Verständnis der Prozesse, die bei optischen Antennen bei linearer und nichtlinearer Anregung auftreten und für ihren Einsatz beispielsweise als Sensoren oder nanoskalige Sender und Empfänger von grundlegender Bedeutung sind. Sie bereiten den Weg für weiterführende Anwendungen, wie beispielsweise die Kombination resonanter Antennenstrukturen mit Quantenpunkten oder der Verwendung zur Erhöhung der Sensitivität von Einzelphotonendetektoren.

Abstract

Optical antennas are metallic nanostructures showing resonance behaviour at optical wavelengths. Similar to macroscopic antennas as used for example in radio wave transmission, their reception and emission properties depend on the size of the structure. More detailed studies show, however, that there are significant deviations from radio frequency antenna behaviour and additional influences. A main reason for this is that metals cease to be perfect conductors at optical frequencies, which makes the assumption of an essentially electromagnetic field-free antenna interior invalid.

In this work, the resonance behaviour of optical antennas made of gold and aluminum is investigated. Specifically, it is of interest how the emission behaviour is changed by the coupling of two antenna arms via a small gap. The influences of the two materials chosen on the antenna performance are also evaluated.

The fabrication process for optical antennas based on electron beam lithography and subsequent evaporation of gold or aluminum is introduced. The topographical properties of the engineered nanostructures with minimal dimensions of about 20 nm are evaluated extensively using electron microscopy and atomic force microscopy.

A measurement setup for optical characterization is developed and its numerous capabilities are presented. Special attention is given to the possibility to apply dark-field microscopy and confocal microscopy. To demonstrate the versatility and flexibility of the setup, the system point spread function for two-photon excitation is measured using a single quantum dot as a probe. In addition, the single photon emission behaviour of quantum dots (antibunching) under the same excitation conditions is demonstrated.

Using dark-field spectroscopy, linear scattering spectra of nanoantennas made of gold or aluminum are measured. The results show that both materials can be used for optical antennas. It is shown as well that the resonance wavelength of aluminum or gold antennas scales linearly with antenna arm length. Two-arm antennas with gap additionally show a spectral red-shift compared to single-arm antennas. For gold, it is experimentally demonstrated that two-arm antennas with gap have superior emission capabilities compared to single-arm antennas. Using gold antennas, the smallest resonance wavelength reached is about 580 nm. Aluminum antennas can be applied to yield even smaller resonance wavelengths. However, these antennas are limited by interband transitions when wavelengths above 700 nm are to be reached. In the wavelength regime where both materials can be used, a larger arm length is required for aluminum antennas. Also, aluminum antennas show a broadened full width at half-maximum compared to gold structures. Their quality factor is lower as well.

Gold and aluminum antennas are also excited using pulsed laser light of wavelength 810 nm, which causes nonlinear effects to occur. For suitably selected excitation intensities, both gold two-arm and single-arm nanoantennas show a plasmonic emission due to two-photon absorption. The spectra of this emission are very similar to linear scattering spectra, even though monochromatic excitation is applied. This two-photon luminescence is explained via excited electronic states in the metal, which relax radiatively under generation of a particle plasmon. For aluminum, very similar effects are observed. This experimental study has both fundamental implications as well as applications. For example, antennas can be characterized quickly for the resonance properties for a particular excitation wavelength, as the emitted intensity scales with the matching of the antenna arm length towards the wavelength used for excitation.

The near-field of two-arm antennas with gap is examined applying a newly developed apertureless scattering near-field optical microscopy method. It is shown that the near-field intensity is significantly enhanced in the antenna gap and also to some extent at the rod ends at the outside of the structure.

An emission response significantly below the diffraction limit is obtained from two-photon excitation of non-resonant single-arm structures via the so-called lightning rod effect. The high concentration of electromagnetic fields is additionally verified using quantum dots as local probes. The single-arm structures are also deformed at the areas of highest field concentration, which is mapped out using atomic force microscopy.

In summary, the effects observed lead to a greater understanding of the processes observed for optical antennas under linear and nonlinear excitation, which are of fundamental interest for their use as sensors or nanoscale receivers and emitters. They pave the way for future applications, such as the combination of resonant antenna structures with single quantum dots or their use to improve the sensitivity of single photon detectors.

Publications

Journal publications

- **M.D. Wissert**, B. Rudat, U. Lemmer, H.-J. Eisler, *Quantum Dots as Single-Photon Sources: Antibunching via Two-Photon Excitation*, **Phys. Rev. B** 83, 113304 (2011).
- **M.D. Wissert**, C. Moosmann, K.S. Ilin, M. Siegel, U. Lemmer, H.-J. Eisler, *Gold Nanoantenna Resonance Diagnostics via Transversal Particle Plasmon Luminescence*, **Optics Express** 19, 3686 (2011).
- **M.D. Wissert**, K.S. Ilin, M. Siegel, U. Lemmer, H.-J. Eisler, *Coupled Nanoantenna Plasmon Resonance Spectra from Two-Photon Laser Excitation*, **Nano Letters** 10, 4161 (2010).
- **M.D. Wissert**, K.S. Ilin, M. Siegel, U. Lemmer, H.-J. Eisler, *Highly Localized Non-Linear Optical White-Light Response at Nanorod Ends from Non-Resonant Excitation*, **Nanoscale** 2, 1018 (2010).
- **M.D. Wissert**, A.W. Schell, K.S. Ilin, M. Siegel, H.-J. Eisler, *Nanoengineering and Characterization of Gold Dipole Nanoantennas with Enhanced Integrated Scattering Properties*, **Nanotechnology** 20, 425203 (2009).
- B. Rudat, E. Birtalan, I. Thomé, D.K. Kölmel, V.L. Horhoiu, **M.D. Wissert**, U. Lemmer, H.-J. Eisler, T.S. Balaban, S. Bräse, *Novel Pyridinium Dyes That Enable Investigations of Peptoids at the Single-Molecule Level*, **J. Phys. Chem. B** 114, 13473 (2010).

Conference proceedings, posters and presentations

- **M.D. Wissert**, C. Moosmann, K.S. Ilin, M. Siegel, U. Lemmer, H.-J. Eisler, *Longitudinal and Transversal Optical Antenna Plasmon Reso-*

nance Spectra from Two-Photon Laser Excitation, **CLEO/Europe EJ4.5**, Munich, Germany, May 22 - May 26, 2011.

- **M.D. Wissert**, C. Moosmann, K.S. Ilin, M. Siegel, U. Lemmer, H.-J. Eisler, *Coupled Nanoantenna Plasmon Resonance Spectra from Two-Photon Laser Excitation: Longitudinal and Transversal Emission*, **DPG Spring Meeting 2011 O11.2**, Dresden, Germany, March 13 - March 18, 2011.

- **M.D. Wissert**, G. Varga, C. Moosmann, K.S. Ilin, M. Siegel, U. Lemmer, H.-J. Eisler, *A White-Light Apertureless Scanning Near-field Optical Microscopy Method for Gold Nanoantennas*, **MRS Fall 2010 M5.28**, Boston, USA, November 30 - December 2, 2010.

- **M.D. Wissert**, K.S. Ilin, M. Siegel, U. Lemmer, H.-J. Eisler, *Longitudinal and Transversal Coupled Nanoantenna Plasmon Resonance Spectra from Two-Photon Laser Excitation*, **MRS Fall 2010 M10.29**, Boston, USA, November 30 - December 2, 2010.

- **M.D. Wissert**, *Engineering of Optical Antennas – Linear and Nonlinear Optical Responses*, **KSOP PhD Seminar**, Strasbourg, France, November 24, 2010.

- **M.D. Wissert**, A.W. Schell, K.S. Ilin, M. Siegel, U. Lemmer, H.-J. Eisler, *Engineering and Characterization of Resonant Optical Antennas*, **MRS Spring 2010 D8.4**, San Francisco, USA, April 5 - April 9, 2010.

- **M.D. Wissert**, *Engineering of Optical Antennas – On the Road to a Tunable Super Emitter*, **Karlsruhe Days of Optics and Photonics**, Karlsruhe, Germany, September 30, 2009.

- **M.D. Wissert**, A.W. Schell, K.S. Ilin, M. Siegel, H.-J. Eisler, *Dark Field Spectroscopy of Optical Antennas*, **Summer School on Plasmonics**, Porquerolles, France, September 13 - September 17, 2009.

- **M.D. Wissert**, *Dark-Field Microscopy and Two-Photon Induced Luminescence*, **KSOP PhD Seminar**, Karlsruhe, Germany, April 29, 2009.

- **M.D. Wissert**, *Optical Antennas*, **KSOP Autumn Colloquium 2008**, Stuttgart, Germany, November 10 - November 11, 2008.

- **M.D. Wissert**, *Engineering of Optical Antennas – On the Road to a Tunable Super Emitter*, **KSOP Summer School**, Bad Herrenalb, Germany, August 20 - August 21, 2008.

Supervision of students

- **P. Schwab**, *Optische Antennen aus Aluminium*, Diplomarbeit (2011).
- **F. Herberger**, *Nanomanipulation von Goldkolloiden*, Studienarbeit (2011).
- **W. Truong**, *Entwicklung und Aufbau eines Konfokaldunkelfeldmikroskopiesystems unter Verwendung kegelförmiger Linsen*, Bachelorarbeit (2010).
- **E. Schmidt**, *Planare optische Antennen aus Aluminium*, Bachelorarbeit (2010).
- **P. Schwab**, *Herstellung und Charakterisierung optischer Quadrupol-Antennen*, Studienarbeit (2010).
- **G. Varga**, *Nahfeldcharakterisierung optischer Antennen*, Diplomarbeit (2009).

Contents

Chapter 1

Introduction

In the last century, antennas have become one of the most important constituents of daily life. In fact, with their ever shrinking size, which makes them easy to hide in many devices, they have probably influenced our existence in more ways than most people realize. From the introduction of the radio to almost any household in the 1920s and 1930s to today's communication possibilities due to mobile phones and smartphones, the sending and receiving of electromagnetic waves using antennas has been a significant part of world history. Optics, most easily defined as the field of science dealing with the properties and the interaction of matter with light, which is also readily described as electromagnetic radiation, has been a similarly important field of research, even for a longer period of time, and also heavily influenced human life, with the laser being the probably most influential recent invention.

Why is it tempting to bring together those two fields of research? Why did the first small ideas of such a combination of concepts in the late 1990s and the first experimental results in the first decade of the 21st century spark an exponentially increasing amount of scientific publications? Most importantly, the reason lies not in long distance communication via light (which would be too easily absorbed in the atmosphere), but in the most interesting confinement of light possible in very small dimensions. Antennas can be understood as efficient transformers from electromagnetic near-fields to far-fields and vice versa, and with optical frequencies, such near-fields and thus light can be confined to a very small volume. In a world which demands for ever faster, smaller, less energy consuming and thus more efficient computing, the confinement of light to very small volumes plays an increasingly important role. Purely optical circuits, consisting of receivers, propagation lines and emitters all designed exclusively for optical frequencies with no need for conversion into electronic states, could provide improved computational capabilities. Even more so as quantum computing, the use of single quantum states as so-called quantum bits (q-bits) in calculations, provides fast and powerful algorithms. Optical antennas in conjunction with single photon emitters, for example quantum dots, may provide photonic q-bits. Optical sensor performance can also be increased using such antennas, where they provide highly localized fields to efficiently probe changes in the optical response of a specific substance.

In this thesis, the engineering of functional optical antennas and extensive characterization techniques will be presented. In the first chapters, the theoretical foundations will be laid out by describing and defining the most important elements and properties of classical antenna theory. Then, the parameters and physical effects influencing the response of optical antennas will be described, highlighting similarities as well as differences to the radio wave counterparts. A short chapter subsequently outlines the luminescence properties of semiconductor quantum dots, which can be beneficially combined with optical antenna structures. The experimental section is headed by the introduction of the antenna fabrication process for gold and aluminum nanoantennas, where a reproducible scheme for nanoantenna engineering is presented.

The characterization part begins with Chapter 5, which details the experimental setup devised for the investigations. Its numerous measuring capabilities are highlighted, focusing on the experimental techniques used throughout the work: dark-field microscopy and confocal microscopy. The measurement of antibunching behaviour for a single quantum dot under two-photon excitation is also presented in this section.

The subsequent chapters each give insight into a specific characterization measurement as applied to optical antennas. Chapter 6 presents extensive optical antenna characterization results using dark-field microscopy, giving insight into the linear response properties of such structures. Main findings are redshifts in emission spectra with increasing antenna size or decreasing two-arm antenna gap size, and significant differences in resonance energy and spectral width between gold and aluminum antennas.

Chapter 7 describes confocal two-photon excitation microscopy of nanoantennas, and reveals luminescent plasmonic relaxations from such structures. The linear and nonlinear emission spectra show striking similarities, which is extensively discussed.

In Chapter 8, an apertureless near-field technique is developed and used to gain a greater insight into the near-field behaviour of optical antennas. The near-field intensity is found to be greatly enhanced in the antenna gap, and also, to a lesser extent, at the antenna ends.

Chapter 9 describes field enhancement effects on non-resonant structures, where a strongly localized two-photon luminescence response significantly below the diffraction limit is observed. This local enhancement is also mapped out using quantum dots as local field-strength sensors.

Finally, Chapter 10 forms the end of the thesis, where conclusions are drawn and an extensive outlook towards future applications is given.

Chapter 2

Optical Antennas: Looking at two Faces of the Same Coin

The fundamental study of optical effects has been a longstanding foundation of physics research. As such, it isn't surprising that many terms concerning such effects have in fact been coined by physicists all over the world. On the other hand, antennas have from their conception at the beginning of the 20th century been a very important element of the then rather new branch of electrical engineering. As such, many terms nowadays commonly used in conjunction with such devices are very well known by engineers, but form only a relatively basic part of a typical physicist's training. Thus it isn't surprising to find that optical antennas, which form a bridging element between the two disciplines, are described with a great variety of terms. In fact, many features blend together for these unique structures, such that it is important to revisit the formalisms typically used by both engineers or physicists and try to make sure that a description can be found that not only most scientists can feel comfortable with, but also to find out which terms need to be adopted to make that description as accurate and nonetheless as simple as possible.

First, however, a definition for an *optical antenna* is required. In the context of this work, an optical antenna is a nanoscopic device converting freely propagating electromagnetic fields at optical frequencies into energy localized in a volume smaller than the antenna itself, or reciprocally converting such localized energy into freely propagating waves. This is analogous to antennas operating at radio-frequencies if for example the current produced in a receiving antenna and used in some device is seen as such a localization. In the characterization steps that are outlined, often a conjunction of processes will be observed, that is the antenna is used both as a receiver and an emitter of optical radiation at the same time.

Optical antenna history has been largely influenced and inspired by classical antenna approaches. After visionary ideas in the 1920s by Synge about localized fields in the vicinity of small particles [174], which were unfortunately largely forgotten during the decades to come, the search for locally enhanced electromagnetic fields was triggered again by phenomena such as surface-enhanced Raman spectroscopy, which was discovered in 1974 by Fleischmann et al. [63] but only explained in 1977 by other groups [88, 3]. Already in 1985, John Wessel noted that a *particle serves as an antenna that receives an incoming electromagnetic field* [186] (see also [16]). Using a gold coated polystyrene particle, Pohl and Fischer experimentally demonstrated these antenna effects for the first time in 1989 [61]. In 1997, bow-tie antennas were applied in the microwave regime by Grober et al. [74], already foreseeing applications at optical frequencies. In 2000, it was again Pohl who described the numerous analogies and possibilities available for optical antennas by considering the successes already made in classical antenna theory [150]. The field has been expanding rapidly ever since the first experimental investigations of optical antennas were published by Mühleschlegel et al. in 2005 [128].

As the experimental results on linear and nonlinear effects in optical antennas presented later in this work will focus on excitation and emission properties of single nanoantenna structures, the main goal of the discussion will be to describe the resonance properties of first classical and then optical antennas, outlining both similarities and differences. This discussion will be followed by the description of nonlinear effects in such structures, which is equally important to the understanding of the two-photon luminescence results shown later. Finally, an overview of possible applications for optical antennas will be given, which is intended to give a feeling for the possibilities to exploit the effects experimentally observed and presented in this thesis.

2.1 Antenna Fundamentals – an Engineer's Point of View

The name antenna for large emission facilities was first introduced by Italian-born Guglielmo Marconi at the end of the nineteenth century, who probably adopted the term from the Italian word for "post" [9]. French scientists soon used the name to describe electromagnetic transmitters [15], and its use has come to be ubiquitous in today's society. The term was finally defined as *a means for radiating or receiving radio waves* by the IEEE in 1983 [7]. Marconi's first antenna to achieve long distance communication, namely in 1901 from Poldhu, Cornwall, England to St. John's, Newfoundland (a British colony at that time), was a fan aerial [9] (see also Figure 2.1), but already at that time several other antenna designs had been proposed and were investigated. Descriptions were quickly formed for numerous designs and properties of such devices.

Figure 2.1: Guglielmo Marconi's antenna in Poldhu, Cornwall, England, as e.g. depicted in [13]

Many antenna designs have been developed over the last century. They are often classified in four main groups – (i) wire type antennas (e.g. long wires, dipoles, and helices), (ii) reflector type antennas (slots, horns, open-ended waveguides), (iii) frequency-independent antennas, that is devices optimized for a large number of wavelengths, and (iv) microstrip/patch antennas (as often found in today's mobile phones). The optical antennas presented in this work fall into the first category, thus only a linear short wire dipole as an example for a wire type antenna will be discussed in greater detail for the remainder of this section.

It was with such a dipole antenna that Heinrich Hertz, then working in Karlsruhe, Germany, conducted the first wireless transmission experiment of any kind in 1886, fifteen years before Marconi's long distance success [39]. A spark in the gap of a half-wavelength dipole (an antenna of a total length of half the wavelength used) was detected in the gap of a second antenna nearby. The wavelength Heinrich Hertz worked with was 4 meters, which is of course far from the optical wavelength regime considered here. Nonetheless, the ex-

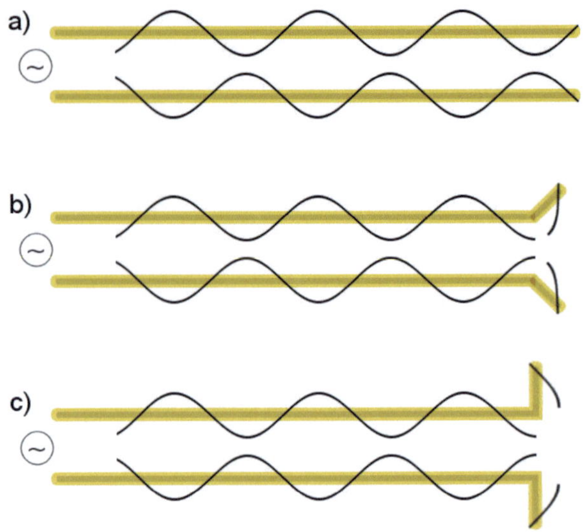

Figure 2.2: a) A two-wire transmission line with a standing wave pattern barely radiates if the two wires are close enough to each other. b) In a flared transmission line, radiation to the far-field becomes significant. c) A linear dipole antenna at the end of a transmission line is an efficient radiator. (Image inspired by Balanis [10])

planation for his experiment also serves well as a first approximation as to at which wavelengths to expect resonances for optical nanoantennas of a specific physical size. How similar the explanations for classical and optical antenna resonances really are will then be discussed in later sections.

In large wavelength (radio frequency, RF) antenna design, one assumes an antenna to be composed of a perfectly conducting metal. In such a metal, charges can be accelerated and decelerated, e.g. by the connection of an AC source to a wire. Such acceleration and deceleration will generate a standing wave pattern on the wire, and also, as Maxwell's equations dictate, electromagnetic radiation. In transmission lines, such a radiation loss is not desired. An AC transmission line is thus typically built using two wires in very close proximity, which are driven with a 180° phase difference, as shown in Figure 2.2 a). The electromagnetic radiation of the standing waves on the two wires then cancels out if the two are close enough to each other, and transmission is maximally efficient. Now imagine that the ends of such a two wire system are slowly bent outwards in opposite direction, as depicted in Figure 2.2 b). Suddenly, some radiation will be able to propagate to the outer world, as the cancellation is no longer complete. If the wire is bent out even more (Figure 2.2 c)), for example, a piece of length $\frac{\lambda}{4}$ for each wire, the interference will suddenly be constructive instead of destructive. Thus the AC driven system will radiate efficiently – a simple antenna has been designed.

Exactly *how* the system will radiate now depends on the exact shape of the wires and the medium the waves propagate into. In the (at least conceptually, probably not from an experimentalist's point of view) easiest case of a very short, perfectly conducting, infinitely thin wire radiating into free space where all feed lines (the wires delivering current to the bent arms) can be neglected, this is just the well-known Hertzian dipole radiation. For more complicated cases, Maxwell's equations need to be solved for the specific boundary conditions of the problem.

2.1.1 Linear Wire Antenna Radiation

As stated before, the question how an antenna radiates is the most important issue for designers of classical antennas. As information is to be transported over great distance, one needs the electromagnetic radiation to propagate in specific directions, as well as to excite that radiation with as few losses as possible. In this section, it will be illustrated in a few steps how this is typically accomplished. This serves to generate an understanding for why antenna language has developed the way it has and even more to show which system parameters are the ones that antenna engineers are mainly concerned with.

A simple linear wire antenna of finite length will be used in the following to describe how an antenna analysis is typically performed and which questions and theoretical requirements arise from there. Of course, this analysis can by no means be complete, as antenna theory is still developing rapidly today, but nonetheless an attempt shall be made to touch on the most important points.

As one of the most simple antenna geometries, a dipole antenna can be thought of as a perfectly conducting small wire of total length L and negligible thickness. To operate as a sending antenna, some current needs to be supplied to that antenna as already seen with the transmission lines in Figure 2.2. This time however, to keep things simple, the feed lines will be neglected and only a current is assumed to be fed to the center of the wire, as depicted in Figure 2.3. Also, the gap between the two antenna arms is assumed to be negligible compared to the total length, such that the results are the same as for a continuous wire of total length L. If the current provided at the feed gap is sinusoidal and the wire is indeed very thin, a sinusoidal current distribution will be generated on the wire as well (neglecting radiation damping). If we assume the wire to be oriented along the z-axis and its center to be at 0, with

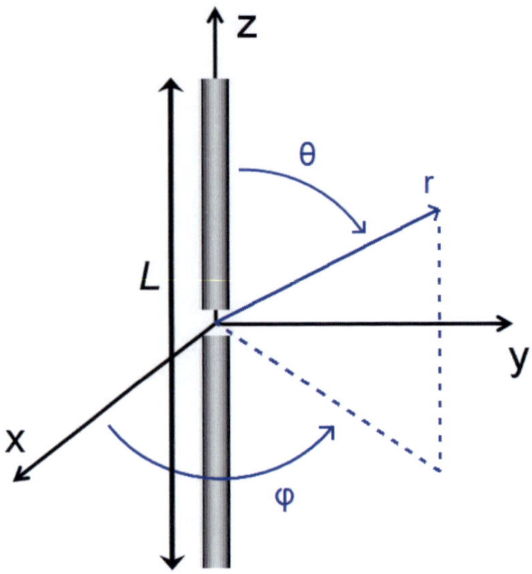

Figure 2.3: Conceptual linear antenna of length L without feed lines

an infinitesimally small gap, then the current density on the wire of total length L can be described as [87]

$$\mathbf{j}(\mathbf{r}) = I \sin\left(\frac{\omega L}{2c} - \frac{\omega}{c}|z|\right)\delta(x)\delta(y)\mathbf{e}_z, \tag{2.1}$$

for $|z| < L/2$, and where I is the maximum current that can be fed to the antenna if $\frac{\omega L}{c} \geq \pi$. ω is the driving frequency of the antenna, while c describes the speed of light in free space.

As described previously, there will now be some destructive and constructive interference of different electromagnetic terms that will lead to propagating electromagnetic waves that are emitted. To find a description for those waves, one must now take a look at the electromagnetic boundary problem that is imposed.

To do this, as a first step, the time convention for electromagnetic fields used throughout the text should be given. Here, the time dependency of the field of a specific angular frequency $\omega = 2\pi f$ is defined to be $\exp(-j\omega t)$, where j is used to describe the imaginary unit. Once this time convention is in place and every field is considered as a superposition of monochromatic plane waves, the field can be written as [139]

$$\mathbf{E}(\mathbf{r}, t) = \mathrm{Re}\{\mathbf{E}(\mathbf{r})\exp(-j\omega t)\}, \tag{2.2}$$

where $\mathbf{E}(\mathbf{r})$ defines the (complex) amplitude of the wave of angular frequency ω. The same convention of course also holds true for the magnetic field component.

Just as any electromagnetic problem, the dipole antenna radiation also needs to fulfill Maxwell's equations, which can be displayed as follows in their differential form for a lossless medium and a time-harmonic field as defined above:

$$\nabla \times \mathbf{E}(\mathbf{r}) = j\omega \mathbf{B}(\mathbf{r}), \tag{2.3}$$

$$\nabla \times \mathbf{H}(\mathbf{r}) = -j\omega \mathbf{D}(\mathbf{r}) + \mathbf{j}(\mathbf{r}), \tag{2.4}$$

$$\nabla \cdot \mathbf{D}(\mathbf{r}) = \rho(\mathbf{r}), \tag{2.5}$$

$$\nabla \cdot \mathbf{B}(\mathbf{r}) = 0. \tag{2.6}$$

Here, **E** and **D** describe the electric field and the electric displacement field, while **H** and **B** are the magnetic field and the magnetic induction. ρ is the total free (electric) charge density, and **j** describes the total free (electric) current density.[1]

Magnetic current and charge densities have not been included, since those quantities do not seem to appear in nature. In antenna theory, however, they are sometimes also present in the equations as they allow for mathematical simplifications of more complicated boundary problems such as for horn antennas [10].

The continuity equation relates current and charge density:

$$\nabla \cdot \mathbf{j}(\mathbf{r}) = j\omega\rho(\mathbf{r}). \tag{2.7}$$

In principle, with those equations, any antenna radiation problem can be solved: Once all source terms, be they electric or conceptually magnetic, are known, the fields at any point in space and time can be found by extensive integration. Unfortunately, this integration is rather complicated and, even with today's computational power, hard to solve for many types of antennas. To circumvent this difficult computation, additional quantities are defined, a vector potential **A** and a scalar potential Φ, which often lead to great simplifications in calculation. These potentials are defined via

[1] In fact Maxwell's equations only give a complete description of what will happen with the fields if additionally the definition for **D** and **B** as well as the so-called *constitutive relations* are given. Without going too much into detail, as it shall not be required for the understanding of the section, it should be noted that $\mathbf{D} = \epsilon_0\mathbf{E} + \mathbf{P}$, where **P** is the electric polarization, and $\mathbf{H} = \frac{\mathbf{B}}{\mu_0} - \mathbf{M}$, where **M** is the magnetization, and the constitutive relations define the (phenomenological) material parameters σ (the conductivity), μ (the permeability), and χ_e (the electric susceptibility): $\mathbf{j} = \sigma\mathbf{E}$, $\mathbf{B} = \mu\mathbf{H}$, and $\mathbf{P} = \epsilon_0\chi_e\mathbf{E}$ (assuming a linear and isotropic medium). [21]

$$\mathbf{B} = \nabla \times \mathbf{A} \tag{2.8}$$

and

$$\mathbf{E} = -\nabla\Phi - \frac{\partial \mathbf{A}}{\partial t}. \tag{2.9}$$

The potentials also need to be found via integration, but it is an integration that is often much easier to perform than the previously presented integration directly via Maxwell's equations. Once they are available, the fields can be found by comparably simple differentiation. For this, some gauge for the potentials can be chosen (that is, an additional constraint is put on the potentials that they are required to fulfill). The Lorenz gauge,

$$\nabla \cdot \mathbf{A} + \frac{1}{c^2}\frac{\partial \Phi}{\partial t} = 0, \tag{2.10}$$

has the great advantage to lead to an uncoupling of the potential equations, yielding two inhomogeneous wave equations, namely (in free space)

$$\nabla^2\Phi - \frac{1}{c^2}\frac{\partial^2\Phi}{\partial^2 t} = -\frac{\rho}{\epsilon_0}, \tag{2.11}$$

$$\nabla^2\mathbf{A} - \frac{1}{c^2}\frac{\partial^2\mathbf{A}}{\partial^2 t} = -\mu_0\mathbf{j}. \tag{2.12}$$

In summary, by defining the vector potential and the scalar potential and introdu-cing the Lorenz gauge, the search for the electromagnetic fields directly (via Maxwell's equations) can be replaced by the search for the vector potentials, from which then the fields can be calculated. This second way is often easier, as the integration to obtain the vector potentials contains fewer difficulties. The vector potentials are usually obtained via the use of the Green's function solving the inhomogeneous wave equations for a delta-distribution-like excitation. The scalar potential Φ is calculated as [18]

$$\Phi(\mathbf{r}) = \frac{1}{4\pi\epsilon_0}\int \rho(\mathbf{r}')\frac{e^{j\frac{\omega}{c}|\mathbf{r}-\mathbf{r}'|}}{|\mathbf{r}-\mathbf{r}'|}dV'. \tag{2.13}$$

Often, there are no free charges and thus no scalar potential, which makes $\mathbf{A}$ the even more important quantity to determine. For time-harmonic fields as defined above, it can be obtained very similarly via [87]

$$\mathbf{A}(\mathbf{r}) = \frac{\mu_0}{4\pi}\int \mathbf{j}(\mathbf{r}')\frac{e^{j\frac{\omega}{c}|\mathbf{r}-\mathbf{r}'|}}{|\mathbf{r}-\mathbf{r}'|}dV'. \tag{2.14}$$

No matter which analytical or numerical solution to the problem has been found, one has thus finally arrived at the electromagnetic fields. For the dipole to be considered in this section, the calculations are fairly straightforward using either method. Typically, for a simple dipole linear antenna, which was the problem defined earlier, the vector potential **A** is obtained first, and then, if there are no free charges and fields are time-harmonic as defined above, the magnetic and electric fields can be calculated via [87]

$$\mathbf{H} = \frac{1}{\mu_0} \nabla \times \mathbf{A} \tag{2.15}$$

and

$$\mathbf{E} = \frac{jZ_0}{\omega / c} \nabla \times \mathbf{H}, \tag{2.16}$$

where $Z_0 = \sqrt{\mu_0 / \epsilon_0}$, the impedance of free space, has been defined.

Combining these equations with equations (2.1) and (2.14), one can thus calculate the electric field of the antennas. Unfortunately, the integration can still be fairly complicated if the field is also to be found very near (distance of less than one wavelength) to the structure, that is if a near-field pattern is to be obtained. In addition, the assumption of the validity of the sinusoidal current distribution already implied some errors in the near-field zone. A more exhaustive and more exact derivation by solving the boundary conditions and thus obtaining a more exact current distribution is outlined in Chapter 9.4B of Jackson [87], who also gives further references. For details on the near-field problem of a linear antenna, see also the treatment by McDonald [119].

RF antenna engineers are mostly concerned with communication over large distances. Consequently, a good way to introduce an approximation to be able to come up with analytical solutions to the problems encountered above is to check for simplifications that arise from an observation of the radiation only at great distances from the antenna. In fact, it was realized early on that in the so-called *far-field* region, by definition *that region of the field of an antenna where the angular field distribution is essentially independent of the distance from the antenna* [7], the distance R from the source could be approximated and thus the calculation reduced significantly. For details on this procedure see e.g. Chapter 4 in reference [10].

For $R >> \lambda$, Equation (2.14) can be simplified, and now reads as [87]

$$\mathbf{A}(\mathbf{r}) = \frac{\mu_0}{4\pi} \frac{e^{jkR}}{R} \int \mathbf{j}(\mathbf{r}') e^{-jk\hat{\mathbf{r}}\mathbf{r}'} dV', \tag{2.17}$$

where $\hat{\mathbf{r}}$ is a unit vector in the direction of $\mathbf{r}$.

A combination of equations (2.1), (2.17), 82.15), and (2.16) alongside a conversion to the more appropriate spherical coordinate system as introduced in Figure 2.3 finally gives for the far-field of the center-fed linear wire antenna of length L [10]

$$E_\theta = jZ_0 \frac{Ie^{-jkR}}{2\pi R} \left[\frac{\cos(\frac{kL}{2}\cos\theta) - \cos(\frac{kL}{2})}{\sin\theta} \right] \tag{2.18}$$

and

$$H_\varphi = \frac{E_\theta}{Z_0}. \tag{2.19}$$

All other field components are 0. Several quantities can be calculated from those fields. The most important parameters and definition shall be given in the following. Two aspects are most fundamental for the analysis: i) the definition of a resonant antenna and ii) the antenna radiation pattern.

2.1.2 Antenna Parameters and Definitions

Input Impedance and Antenna Resonance

To model an antenna in circuit theory, it is often easiest to describe it with the use of an equivalent circuit, transforming its properties into lumped elements, namely a complex input impedance with real and imaginary part, easily defined via

$$Z_{\mathrm{ant}} = R_{\mathrm{ant}} + jX_{\mathrm{ant}}. \tag{2.20}$$

The real part R_{ant} consists of the sum of the so-called radiation resistance (which to the person using an antenna are the good thing, as they describe the radiation), R_{rad}, and an unwanted conductive and dielectric resistance responsible for losses, R_{L}, which is 0 for a perfectly conducting antenna.

An antenna in transmitting mode (that is driven via a source and generating radiation), can thus be depicted in a Thévenin equivalent circuit, see Figure 2.4.

If $R_{\mathrm{L}} \neq 0$, one can define a *radiation efficiency* η, via [9]

$$\eta = \frac{R_{\mathrm{rad}}}{R_{\mathrm{rad}} + R_{\mathrm{L}}}. \tag{2.21}$$

As the antenna is connected to the rest of the electrical engineering world via feed lines, for a center-driven antenna the feed gap is the connecting point between the antenna and its driving circuit (red points in Figure 2.4). The

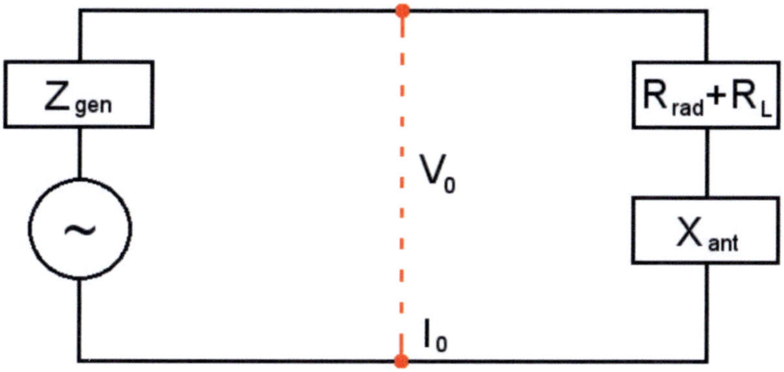

Figure 2.4: A transmitting antenna depicted in a Thévenin equivalent circuit. The left side depicts the source, the antenna is at the right side. At the feed gap (red points), a voltage and a current are present.

resistance can also be measured by applying a voltage V_0 and measuring the current I_0 at the feed gap, so the impedance can be defined as well by

$$Z_{\text{ant}} = \frac{V_0}{I_0}. \tag{2.22}$$

An antenna is said to be *resonant* when this impedance becomes entirely real (and thus, if losses can be neglected, identical to R_{rad}), because it can then be easily matched to a feeding element (often assumed to have an entirely real impedance of 50 Ω). The input impedance can be calculated using the *induced EMF method* [10], and one finds that for a center-driven linear wire antenna of length of approximately $\frac{\lambda}{2}$, the antenna is in such a resonant condition. This is again the case for odd multiples of $\frac{\lambda}{2}$, while the reactance becomes infinite (as seen at the feed gap) for even multiples of $\frac{\lambda}{2}$, which are thus a severely detuned case. This can also be modeled using reflection coefficients, where a reflection coefficient of 0 implies that as much power as possible is transferred to the antenna and a reflection coefficient of 1 means that no power at all is transferred. A plot for the reflection coefficient (in a logarithmic scale) is given in Figure 2.5 for an antenna of 1 m total length for different frequencies. Obviously, the reflection coefficient significantly drops close to 150 MHz and 450 MHz, which are the cases where the antenna constitutes an element of length $\frac{\lambda}{2}$ and $\frac{3\lambda}{2}$.

It should be noted that, speaking in the Thévenin equivalent circuit model, any wire length can be made resonant by an appropriate addition of capacitance or reactance, again balancing the antenna towards the load (i.e. the feed

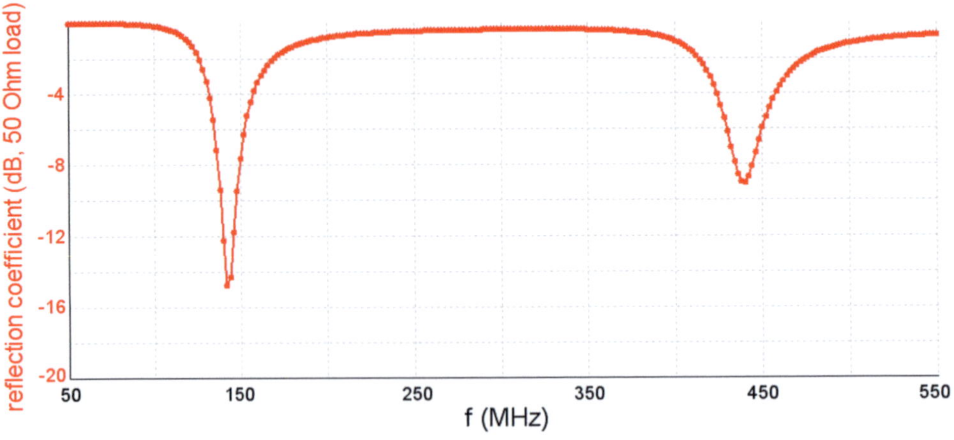

Figure 2.5: Logarithmic reflection coefficient for a center-driven thin wire antenna of 1 m length in free space. Resonances are clearly visible near 150 MHz and 450 MHz. Data generated using freeware software 4nec2.

line and generator in transmission mode). The basic $\frac{\lambda}{2}$ antenna is preferred against its odd multiple counterparts (and thus becomes the typical textbook antenna) because the real part of its resistance R_{ant} is 42.5 Ω (see e.g. Chapter 8 of reference [10]), which can be conveniently connected without large mismatch losses to a typical coaxial cable of resistance 50 Ω. Additionally, its radiation pattern is much preferrable to longer antennas, which shall be disussed in the next section.

Radiation Pattern and Directivity

From the fields calculated in Equation (2.18) and (2.19), the total power density of the radiation (in units of $\frac{\text{W}}{\text{m}^2}$) can be defined. It can be calculated using the complex Poynting vector,

$$\mathbf{S} = \frac{1}{2}\mathbf{E} \times \mathbf{H}^* = \mathbf{S}_r + j\mathbf{S}_i, \tag{2.23}$$

of which the imaginary part is dominant in the near-field and the real part dominant in the far-field [9].

From this, three crucial antenna parameters are derived. As the first quantity, the *radiation intensity* U is the power radiated from an antenna per unit solid angle,

$$U = R^2 S_r, \tag{2.24}$$

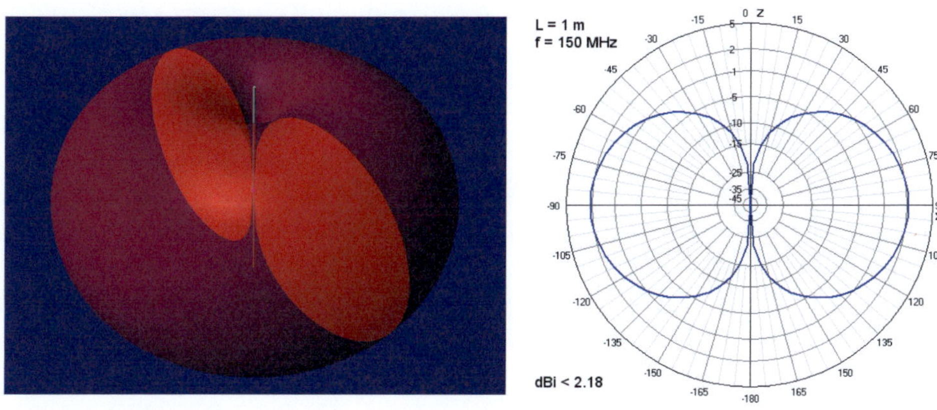

Figure 2.6: 3D directivity for a half-wave antenna (left), and central cut along the long antenna axis (right) as exemplified by a 1 m length structure driven at 150 MHz. Data generated using freeware software 4nec2.

which for the linear dipole discussed above is calculated as [87]

$$U_{\text{dip}} = \frac{Z_0 I^2}{8\pi^2} \left| \frac{\cos(\frac{kL}{2}\cos\theta) - \cos(\frac{kL}{2})}{\sin\theta} \right|^2. \tag{2.25}$$

When this quantity is normalized to the total emitted antenna power P_r, the *directivity* D in a specific direction is obtained:

$$D(\theta,\varphi) = 4\pi \frac{U(\theta,\varphi)}{P_r}. \tag{2.26}$$

Typically, if no other statement is made, the directivity is given for the direction in which most power is directed. For example, the (maximum) directivity of a half-wavelength linear antenna ($L = \frac{\lambda}{2}$) is 1.67 [9], which implies that it emits 1.67 times as strong in its maximum direction (which is for $\theta = 0$) compared to as if it were emitting isotropically. In exchange, its directivity is 0 for $\theta = \frac{\pi}{2}$. Figure 2.6 gives an exemplary radiation pattern for the 1 m length antenna already used as an example above, driven at 150 MHz, the frequency for which it is a half-wave antenna. The directivity is shown in a logarithmic plot in dBi, where $D_{\text{dBi}} = 10\log D$.

For an antenna with losses, i.e. $R_{\text{L}} \neq 0$, some power is lost as we have seen previously, so for an antenna that is receiving, one may also define an antenna *gain* G via

$$G(\theta,\varphi) = \eta \cdot D(\theta,\varphi), \tag{2.27}$$

where η is the radiation efficiency as defined in Equation (2.21). For a lossless antenna, directivity and gain are identical.

Reciprocity Theorem

The ultimate goal of anyone using a classical RF antenna is to send information from some point A (the current whereabouts of Alice) to another point B (where Bob is located) and vice versa. Should Alice and Bob, who both require a transmitting and a receiving antenna, buy two antennas each, or does one suffice? The antenna reciprocity theorem conveniently gives answers to this logistical predicament.

As before only considering electrical currents, the reciprocity theorem may be written as [10]

$$\int \mathbf{E}_2 \cdot \mathbf{j}_1 dV_1 = \int \mathbf{E}_1 \cdot \mathbf{j}_2 dV_2, \tag{2.28}$$

where *reactions* between sources and fields have been defined. If a current in region 1 reacts with fields emanating from region 2, this is the same as the reaction of a current in region 2 with fields from region 1. Reciprocity has two important consequences for antenna properties. First, it can be shown that the power transmitted from A to B is the same as the power transmitted from B to A if the same amount of power is input into the sending antenna. Maybe even more important for the remainder of this thesis is the fact that the emission pattern of the antenna equals its receiving pattern. Thus once one of the two is known, the second is available right away. For example, it can directly be concluded that a linear antenna is best excited if electromagnetic radiation comes in along its maximum directivity. Both properties are derived e.g. in references [10, 37].

Polarization Dependency and Effective Area

One last important property of electromagnetic fields has been neglected so far and should be mentioned at this point – polarization. For a linear wire antenna, the radiation emitted can be considered to be linearly polarized along the long antenna axis in the far-field. If a second linear antenna is supposed to receive some of the emitted radiation, it would be best if it were oriented the same way as the original antenna. If not, some power that otherwise could be received, will be lost. To describe that mismatch, a *polarization loss factor* p_{lf} can be introduced. As it is defined for the power that can still be used compared to the ideal case of matched polarization, it reads

$$p_{\mathrm{lf}} = \cos^2\alpha, \tag{2.29}$$

where α describes the angle mismatch between the two antennas. For many transmission systems, circularly polarized radiation is emitted by the transmitter, while the receivers detect linearly. With this setup, only 50% can be received – but reception won't drop to zero for any orientation of the receiving element.

Using this knowledge, one can finally define an *antenna effective area* [15] A_{e} via

$$A_{\mathrm{e}}(\theta,\varphi,\mathbf{n}_{\mathrm{pol}}) = \frac{P_{\mathrm{coll}}}{I(\theta,\varphi,\mathbf{n}_{\mathrm{pol}})}, \tag{2.30}$$

where I is the incident intensity from a certain angle and with a certain polarization, for which then the efficient area or antenna aperture is defined from the collection of a certain power P_{coll}. Since this is the power that remains within the antenna-receiver system, the definition is the same as for an absorption cross section, which will be encountered again a few chapters later.

Towards Optical Antennas

In summary, it has been shown that classical RF antenna engineers are typically not concerned with material properties. The antenna is assumed to be perfectly conducting, and this can be achieved for typical wavelengths with many metals. So their attention is clearly aimed at the antenna shape, which is used to form propagating fields in the far-field. Most importantly, in specific directions, antenna radiation can be enhanced significantly, such that this direction becomes the preferred direction of emission or receiving. Thus the radiation patterns of such an antenna need to be considered. The impedance matching towards connecting circuits is of equal importance. Only unavoidable losses are allowed to occur to make sure that the antenna is used efficiently. From all those issues, several terms arise that are important to describe antenna performance, which have been described previously.

Now that classical antennas have been described in great detail, how do optical antennas differ, and, even more importantly, why? A rather obvious difference is the fact that currently no negligible feed lines can be connected to such structures and no AC source is available to operate at optical frequencies. This means that only radiative excitation and responses can be observed. Optical antennas could, however, be driven using a local oscillator. For example, a small quantum dot radiating in the feed gap of a two-arm dipole antenna should be able to couple its emission to the antenna – or *vice versa* be fed by the antenna. The general answer to the question, however, will lead us back to the very first assumption we made when we introduced the dipole antenna – that of a perfectly conducting metal. Unfortunately, all metals cease to have that property in the visible wavelength regime. Electromagnetic fields will therefore penetrate a certain distance into the metal (the so-called *skin depth*), and the material properties, i.e. the real and imaginary part of its dielectric constant, will have an impact on the antenna performance. Due to this difference, even shape variations on a small scale will influence the resonance condition, which is now defined solely by the antenna itself and its nearby environment, as coupling to a circuit is no longer required. It should be noted at this point already that, while Maxwell's equations are inherently scale invariant, the resonance conditions are not, as they are also heavily influenced by material properties and thus a changing dielectric function ϵ. All these influences can be dealt with using the well established theory of localized surface plasmons (often also termed particle plasmons), which will be introduced in the next section.

2.2 Localized Surface Plasmons

The optical antennas investigated experimentally in later chapters are fabricated using gold and aluminum. In a first step, it is thus required to gather as much information as possible about the substance they are made of. The materials commonly known as *metals* consist of atoms that, when they form a bulk substance, easily become ionized such that many free electrons are available within the bulk. These free electrons, which form the so-called metallic bonding, give rise to several typical metallic properties such as their high conductivity.

When interacting with light of the right frequency, the free electrons can start to oscillate against the free ions. Those oscillations are typically termed plasmons, where a single plasmon is the description for a quantized electron oscillation. When the electrons are also coupled to a photon, one should strictly speak of plasmon polaritons. In practical usage, the polariton is often dropped, so plasmons can refer to both simple plasmons or plasmon polaritons.

To achieve a quantitative description of the electronic oscillation, one first needs to come up with a model for the so-called free electrons. Depending on just how *free* those really are (that is, the amount and type of restoring forces available), there are several models that have been developed to describe them. It is thus the intent to find the dielectric function $\epsilon(\omega)$ that best describes the properties of the light-matter interaction for a metallic, non-magnetic ($\mu = 1$) material.

Once a proper description for the metal has been found, the following sections will deal with different ways to come up with a resonance condition for metallic nanoparticles, specifically optical antennas: classical Mie theory (as described e.g. in [21]), electrostatic approximations, and finally Fabry-Pérot-type effective wavelength scaling as recently suggested by Novotny [137], which will enable an easier comparison with the classical large wavelength antennas presented earlier.

2.2.1 Drude-Sommerfeld-Model

Before metal properties can be explained, an even more simple concept has to be quickly revisited. As we have already seen in the section on classical antennas, the response of a material to electromagnetic waves crucially depends on its dielectric and magnetic properties. If the latter are assumed to be negligible ($\mu = 1$), then the dielectric constant ϵ plays the major role. The fact that it is often called a *constant* already implies the main difference that will now be encountered for optical frequencies. At higher energies, ϵ is highly frequency dependent and thus considered a dielectric function.

When looking at this dielectric function $\epsilon(\omega)$ from a macroscopic point of view, it is most easily defined by the polarization $\mathbf{P}(\omega)$. For an isotropic and linear material, this polarization is connected to the incident field by the electric susceptibility $\chi_\mathrm{e}(\omega)$ via

$$\mathbf{P}(\omega) = \epsilon_0 \chi_\mathrm{e}(\omega)\mathbf{E}(\omega). \tag{2.31}$$

At the same time, the polarization is also defined by the sum of all N small dipole moments $\mathbf{m}_\mathrm{dip}$ that microscopically add up per unit volume:

$$\mathbf{P}(\omega) = \sum_{i=1}^{N} \mathbf{m}_\mathrm{dip,i}(\omega). \tag{2.32}$$

If only free electrons of charge e are present, this can be further simplified to yield

$$\mathbf{P}(\omega) = N\mathbf{m}_\mathrm{dip} = Ne\mathbf{r}(\omega), \tag{2.33}$$

where $\mathbf{r}$ describes the displacement of an electron from its equilibrium position due to an incident electromagnetic field. If the displacement $\mathbf{r}$ is known, the polarization $\mathbf{P}(\omega)$ and thus also χ_e can be calculated.

As the polarization is also defined via the electric displacement **D**,

$$\mathbf{D}(\omega) = \epsilon_0\epsilon(\omega)\mathbf{E}(\omega) = \epsilon_0\mathbf{E}(\omega) + \mathbf{P}(\omega), \tag{2.34}$$

the (generally complex) dielectric function can be calculated from χ_e and thus from $\mathbf{r}$ by using[2]

$$\epsilon(\omega) = 1 + \chi_e(\omega). \tag{2.35}$$

In the most simple picture for the dielectric properties of a metal, the *Drude model*, all electrons are treated as point-like solid spheres freely moving around fixed, larger and equally solid spheres (the ions) within the bulk material. The mean free movement of the electrons is generated from random collisions of electrons with ions after a statistical time τ_{el} (electrons don't collide with each other), thus their mean speed can be described by the Boltzmann theory. While this simple model is able to generate some valid results (and falls short in other areas, such as heat conductivity), it was greatly improved by some additional influences introduced by Sommerfeld. He took into account an effective electron mass m_{eff}, which was different from the mass of a free electron and conceptually incorporated influences from the interaction potential of the ion grid with the sea of electrons. Additionally, Sommerfeld introduced the Fermi-Dirac-distribution instead of the Boltzmann-distribution for the speed of electrons, thereby accounting for their quantum mechanical properties (i.e. that the Pauli principle has to be obeyed), considerations which are based on the symmetry of the total wavefunction for fermions or bosons. Some later refinements of the Drude-Sommerfeld-model also added corrections for the band structure of the respective materials (albeit only intraband transitions were included). In the Drude-Sommerfeld-model, the electrons of effective mass m_{eff} and elementary charge e thus respond to an external field of amplitude $\mathbf{E}_0$ and frequency ω. Their damping factor is γ, which incorporates the Fermi-velocity v_F and the mean free electron path l_m, $\gamma = \frac{v_F}{l_m}$. The differential equation describing the electron behaviour thus takes the form of a damped Lorentzian oscillator without restoring force (as the electrons are free):

$$m_{\text{eff}}\frac{\partial^2\mathbf{r}}{\partial t^2} + m_{\text{eff}}\gamma\frac{\partial\mathbf{r}}{\partial t} = e\mathbf{E}_0 e^{-j\omega t} \tag{2.36}$$

When assuming that the solution for the position $\mathbf{r}$ of the electron has the same time harmonic behaviour as the incident field, one can thus calculate a

[2] With this point of view, the conductivity does not form a part of the dielectric function. To only require to work with a single quantity including all processes, one can define [139] a complex dielectric function also incorporating energy dissipation from conduction for lossy media as $\epsilon(\omega) = 1 + \chi_e(\omega) + j\frac{\sigma}{\omega\epsilon_0}$.

material	$\hbar\omega_{\mathrm{p}}$	γ	ϵ_∞
gold	9.03 eV	$6.46\cdot10^{12}\frac{1}{\mathrm{s}}$	9.84
aluminum	14.76 eV	$19.79\cdot10^{12}\frac{1}{\mathrm{s}}$	0.2

Table 2.1: Drude-parameters ω_{p} and γ for gold and aluminum [144, 130], as well as a value for the contribution of bound electrons to the dielectric function for gold [170] and aluminum [35, 53].

solution, use it with Equation (2.35) and finally find that

$$\epsilon_{\text{Drude-Sommerfeld}} = 1 - \frac{\frac{ne^2}{m_{\text{eff}}\epsilon_0}}{\omega^2 + j\gamma\omega} =: 1 - \frac{\omega_{\mathrm{p}}^2}{\omega^2 + j\gamma\omega}, \tag{2.37}$$

where ω_{p}, the (bulk) plasma frequency for the material in question, has been defined. Values for gold and aluminum, the materials used later in this work, are given in Table 2.1.

If one now takes a look at measured values for the dielectric function (which are for example measured by observing transmission and reflection of thin films [89]), one typically finds that the behaviour is well-represented using the Drude-Sommerfeld-model for frequencies up to the infrared regime. For even higher energies, bound electrons need to be considered in addition to the free ones. The reason for this behaviour is that high energy photons can transfer enough energy and momentum to such electrons that they are promoted into the conduction band. Conceptually, this may be described as the excitation of oscillations of such bound electrons [139]. From solving the equation of motion for such a Lorentz oscillator, which is described by a relation similar to Equation (2.36), with the addition of a restoring force with spring constant ς, one comes up with the additive interband contribution to the dielectric function [139]

$$\epsilon_{\text{interband}} = 1 + \frac{\frac{n_{\text{bound}}e^2}{m_{\text{eff, bound}}\epsilon_0}}{(\omega_0^2 - \omega^2) - j\gamma\omega} =: 1 + \frac{\omega_{\text{p,bound}}^2}{(\omega_0^2 - \omega^2) - j\gamma\omega}, \tag{2.38}$$

where the bound values are defined as their free-electron counterparts and $\omega_0 = \sqrt{\frac{\varsigma}{m_{\text{eff, bound}}}}$ describes the frequency at which the interband transition occurs. This correction yields satisfactory results for the modeling of bulk dielectric effects, although an additional offset correction with a linear addition

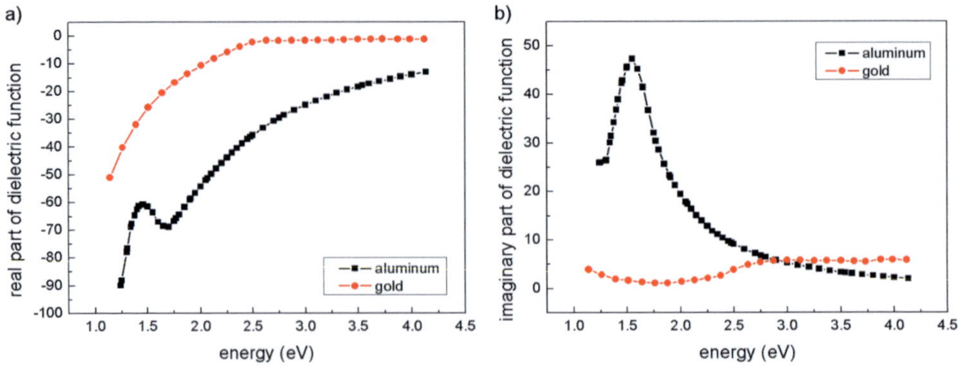

Figure 2.7: a) real and b) imaginary part of the dielectric function for gold and aluminum in the optical regime as measured in references [89, 145].

to ϵ is required to account for all higher-energy interband transitions [170]. This offset is often called ϵ_∞ and is also included in Table 2.1.

Despite all these modeling efforts, which can yield significant insight, it is still best to resort to measured values for the dielectric functions, as for example the fitting parameters and models obtained may greatly vary in the literature. The measured dielectric function of gold and aluminum [89, 145] is plotted in Figure 2.7. A Drude model fit can be performed to obtain satisfactory results for gold. For aluminum, one interband contribution has to be considered additionally. It should be noted at this point that the real part of the dielectric function generally increases with frequency for both gold and aluminum above about 1.75 eV (that is for wavelengths below about 700 nm and thus a large part of the visible spectrum), an important observation that will have consequences later on.

Now that an adequate description for the dielectric function has been found that can be used to model it for a substantial part of the frequency band and yield satisfactory results[3], it is time to check on the resonance condition for single and (later) coupled nanoparticles, which are considered in this thesis as optical antennas. To do this, we will as announced in the introduction to this chapter, first take a look at Mie theory and its implications, then look at classical approaches to find a description of particle resonance, especially spheres

[3]On a side note, the real part and the imaginary part of the electric constant are not independent. The Kramers-Kronig relation (see for example reference [21]) dictates that only vacuum can have no absorption at all frequencies (and that thus all assumptions of lossless media are definitely only just that – assumptions) and that if the real part of the dielectric function is frequency dependent, the imaginary part must be frequency dependent as well (and vice versa). Finally, it must also be true that at very high frequencies at some point the real part of ϵ converges towards unity.

and prolate spheroids, and then in a final step take a look at the resonance of an elongated cylinder as the solution to a waveguide problem.

2.2.2 Absorption and Scattering of Small Particles

To understand the scattering behaviour of arbitrarily shaped particles or particle pairs as used here to design optical antennas, it is helpful to start out with the most symmetric small particle available, a nanoscopic sphere. At the beginning of the 20th century already, several scientists were concerned with the solution to the electromagnetic problem of scattering and absorption by such a particle. An analytical solution was found by Gustav Mie, who published his theory in 1908, which today often bears his name alone, even though he may not have been the first to develop it [21]. Here we shall adopt the name Lorenz-Mie-Debye theory to accomodate for all three researchers nowadays commonly associated with the derivation.

Lorenz-Mie-Debye Theory

Introducing several abbreviations, the main result of Mie's derivation is summarized in the scattering and absorption coefficients he calculates for spherical particles (the derivation can be found in the book by Bohren and Huffman (Chapter 4) [21], from which the equations are adapted):

$$a_n = \frac{m\psi_n(mx)\psi_n'(x) - \psi_n(x)\psi_n'(mx)}{m\psi_n(mx)\xi_n'(x) - \xi_n(x)\psi_n'(mx)}, \tag{2.39}$$

$$b_n = \frac{\psi_n(mx)\psi_n'(x) - m\psi_n(x)\psi_n'(mx)}{\psi_n(mx)\xi_n'(x) - m\xi_n(x)\psi_n'(mx)}, \tag{2.40}$$

where ψ and ξ stand for Riccati-Bessel functions, m is a convenient abbreviation for the square root of the relative complex dielectric function of the system ($m = \sqrt{\frac{\epsilon_{\text{particle}}}{\epsilon_{\text{med}}}}$), and $x = \frac{2\pi\sqrt{\epsilon_{\text{med}}}R}{\lambda}$ is a size parameter combining several other system properties including the particle radius R and the free space wavelength λ. One notes that the coefficients all show an index n. This indexing is due to the fact that, in order to achieve an analytical solution, an inciding plane wave must be expanded into spherical harmonics, thus yielding an infinite series of components.

Those coefficients enter into the respective scattering and absorption sections C_{sca} and C_{abs} for spherical particles, which are found as infinite sums over all plane wave expansion components:

$$C_{sca} = \frac{2\pi}{k^2} \sum_{n=1}^{\infty} (2n+1)(|a_n|^2 + |b_n|^2), \tag{2.41}$$

$$C_{abs} = \frac{2\pi}{k^2} \sum_{n=1}^{\infty} (2n+1)\mathrm{Re}\{a_n + b_n\}, \tag{2.42}$$

where $k = \frac{2\pi \sqrt{\epsilon_{medium}}}{\lambda}$ is the wavevector in the medium in which the sphere is embedded.

Using these equations, which can be implemented into today's computer systems with comparable ease (see e.g. Sönnichsen's PhD thesis for an example [170]), one can thus find scattering resonances for spherical particles of arbitrary size. Two drawbacks limit the general applicability, however. First of all, the theory is inherently limited to spheres, and even though numerous extensions have been presented for particles of other sizes and shapes (such as ellipsoids, which are already closer to the shape of an arm of nanoantennas), they are all not as exact as the original equations for spheres [99]. In any case, approximate solutions for such particles are required, which shall be dealt with in the next sections. Also, while the spectral shapes and shifts are accurately reproduced using Lorenz-Mie-Debye theory, the equations may hardly be used for intuitive guesses to arrive at quick estimates for design rules. The applicability of the equations also crucially depends on the values of the dielectric functions in question. As described in the previous chapter, those can be calculated or measured, but there still remains some uncertainty. With these drawbacks in mind, some of which can hardly be avoided, the next section will introduce electrostatic approaches, which can give faster and more intuitive results for the scattering cross sections if some additional assumptions are made.

Quasi-static Models

A significant simplification of the problem is achieved when one assumes that the particle in question is very small compared to the wavelength of the incident light. This may not be true for all antennas studied subsequently in this work (which can reach a total length of about one third of the respective resonance wavelength), but is still a valid first approximation. Why is this simplification so important? When the antenna is much smaller than the wavelength, then the phase and thus the momentary amplitude of the electromagnetic wave is approximately the same all over the structure, so retardation effects don't have to be taken into account. This means that a quasi-static calculation can be performed (*quasi*-static because the fields still oscillate in time), as charges are simply moved back and forth within the particle, as depicted in Figure 2.8. Of course, this simple picture is only valid if the field within the particle under investigation is in fact homogeneous – something one ordinarily wouldn't expect from metals, as the fields decay exponentially into them. So particles should be smaller than the already mentioned skin depth of the metal for the following assumptions to be completely valid. The skin depth δ for materials at optical frequencies is defined through decay in a medium with an imaginary part of the index of refraction n as [80]

$$\delta = \frac{\lambda}{2\pi \mathrm{Im}\{\mathrm{n}\}}, \tag{2.43}$$

which in the framework of this work is best expressed using real and imaginary part of the dielectric function, yielding

$$\delta = \frac{\lambda}{2\pi}\sqrt{\frac{2}{\sqrt{\mathrm{Re}\{\epsilon\}^2 + \mathrm{Im}\{\epsilon\}^2} - \mathrm{Re}\{\epsilon\}}} \approx \frac{\lambda}{2\pi\sqrt{|\mathrm{Re}\{\epsilon\}|}}, \tag{2.44}$$

where the approximation is valid for negative real parts of epsilon with $|\mathrm{Re}\{\epsilon\}| >> \mathrm{Im}\{\epsilon\}$, which is an acceptable approximation both for gold and aluminum in the optical regime [163, 164, 89] (compare also Figure 2.7). Examples for the skin-depth for gold and aluminum, the materials used in experiments later in this work, are given in Table 2.2. Generally speaking, the quasi-static model is thus a very good model for gold structures of 20 nm width as considered later in this work, but is somewhat limited to predict the behaviour of aluminum structures of the same size.

To calculate the particle resonance of a structure in the quasi-static limit, one needs to solve the Laplace equation for the potential,

$$\nabla^2 \Phi = 0, \tag{2.45}$$

with the appropriate boundary condition for the particle, from which then the electromagnetic field can be deduced via

$$\mathbf{E} = -\nabla \Phi. \tag{2.46}$$

If one performs this calculation, the end result will be that the scattered field is in fact nothing but the electrostatic field of a conceptual dipole located

wavelength/nm	skin depth gold/nm	skin depth aluminum/nm
800	26.3	18.1
700	28.1	15.9
600	32.2	15.9
500	43.2	16.6
450	38.1	16.6

Table 2.2: Skin depths at various optical wavelengths for gold and aluminum. The values were calculated using the imaginary part of the refractive index as measured by Schulz [163].

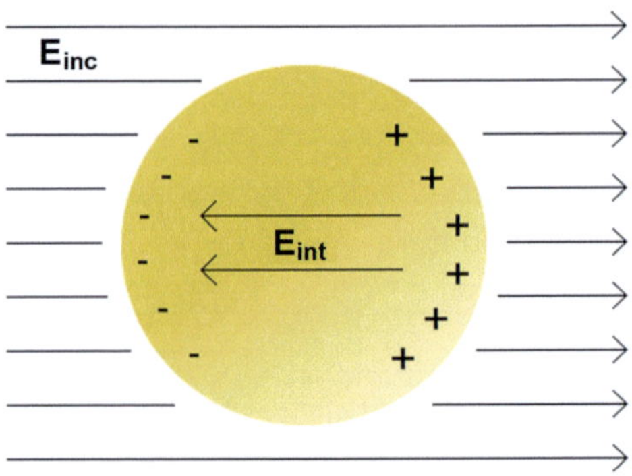

Figure 2.8: Momentary electric fields around a particle small compared to the wavelength

at the center of the spherical particle [139]. The main result is the polarizability α of the particle, which is a measure for the strength of the dipole moment created by the incident field ($\mathbf{m}_{\text{dip}} = \alpha\epsilon_{\text{med}}\mathbf{E}$) and is found to be

$$\alpha(\omega) = 4\pi\epsilon_0 R^3 \frac{\epsilon_{\text{sph}}(\omega) - \epsilon_{\text{med}}}{\epsilon_{\text{sph}}(\omega) + 2\epsilon_{\text{med}}}. \tag{2.47}$$

This polarizability can then be used to find the scattering cross section and the absorption cross section of the sphere. Both are defined as effective areas interacting with the incident light intensity. For the absorption and scattering cross sections, the amount of power absorbed into or scattered by the particle, respectively, is divided by the incident intensity. A similar concept has already been encountered for the effective area of a classical RF antenna in Equation (2.30). Figuratively speaking, the cross sections are thus measures for the apparent size of the particle towards the incident light, which can be more or less than the actual physical extent. They are found to be

$$C_{\mathrm{sca}} = \frac{k^4}{6\pi\epsilon_0^2}|\alpha(\omega)|^2 \tag{2.48}$$

and

$$C_{\mathrm{abs}} = \frac{k}{\epsilon_0}\mathrm{Im}\{\alpha(\omega)\}. \tag{2.49}$$

Considering these results, one finds that the resonance for spherical structures is thus given for the frequency at which $\epsilon_{\mathrm{particle}}(\omega) = -2\epsilon_{\mathrm{med}}$. As the dielectric constant of the medium is typically real, while the dielectric function of the particle consists of a real and an imaginary part, this resonance condition requires that $\mathrm{Re}\{\epsilon_{\mathrm{particle}}(\omega)\} = -2\epsilon_{\mathrm{med}}$ and $\mathrm{Im}\{\epsilon_{\mathrm{particle}}\}(\omega) \approx 0$. The frequency at which this condition is fulfilled is often termed Fröhlich-frequency. This frequency thus determines the dipole particle plasmon resonance.

It is also interesting to note that the absorption cross section scales with the cubed radius (and thus the volume) of the particle, while the scattering scales with the squared volume. In other words,

$$\frac{C_{\mathrm{sca}}}{C_{\mathrm{abs}}} \propto V. \tag{2.50}$$

This directly implies that absorption becomes dominant for small particles and scattering is more important (and thus easier to observe) for larger ones. A typical example for the dependency of the absorption and scattering properties for a specific particle material is the Lycurgus cup exhibited in the British museum, depicted in Figure 2.9. Due to silver-gold alloy nanoparticles (ratio of silver to gold about 7:3, size of about 50 nm), light of reddish colour is predominantly transmitted, while green light is predominantly scattered [11, 65].

While equations (2.48) and (2.49) do yield significant insight into scattering and absorption by spheres, there is one serious drawback – the resonance condition obtained is strictly speaking only valid for vansishingly small radii [21]. This explains one curious feature of the obtained resonance – its spectral position is independent of particle size. Especially since the particles considered later in this work are somewhat too large to justify the assumptions made above, it would be nice to find at least a hint as to in which directions the resonance frequency would shift. In fact, this can be provided by taking again a look at the similarities of Mie theory and the electrostatic approximation: The Fröhlich-resonance condition can also be obtained by only considering the first order expansions of Bessel functions [21]. This directly implies that an intermediate regime for the resonance for small, finite sized particles can be

Figure 2.9: The Lycurgus Cup when photographed without (left) and with (right) the use of a flash. In the first image, the transmission contribution is dominant, in the second, the scattering contribution. (Photographs from a personal visit to the British Museum)

obtained by retaining more terms in the series expansions of the plane waves from Mie theory. One then finds that [21] the next best approximation for the resonance is

$$\epsilon_{\text{particle}} = -\left(2 + \frac{48\pi^2}{5}\frac{\epsilon_{\text{med}}}{\lambda^2}R^2\right)\epsilon_{\text{med}}. \tag{2.51}$$

This gives us one additional information about resonance scaling with particle size: Depending on the slope of the dielectric function around the resonance condition, the particle size will have significant influence on the resonance. Typically, as we have seen in the preceding chapter, the real part of the dielectric function is increasing with frequency around the Fröhlich-frequency, so larger particles will have a resonance shifted towards longer wavelengths (smaller energies). This is a feature that is well reproduced for the experimental data presented later.

More information can be obtained from the polarizability if even more terms of Mie's theory are retained [121, 102, 116]. Then, α contains even higher terms, and an extensive number of predictions can be made. The polarizability so obtained reads

$$\alpha = \frac{1 - \frac{1}{10}(\epsilon_{\text{sph}} + \epsilon_{\text{med}})x^2 + O(x^4)}{(\frac{1}{3} + \frac{\epsilon_{\text{med}}}{\epsilon_{\text{sph}} - \epsilon_{\text{med}}}) - \frac{1}{30}(\epsilon_{\text{sph}} + 10\epsilon_{\text{med}})x^2 - j\frac{16\pi^2 {\epsilon_{\text{medium}}}^{3/2}}{9}\frac{R^3}{\lambda^3} + O(x^4)} \frac{4\pi R^3}{3}, \tag{2.52}$$

where the size parameter $x = \frac{2\pi R}{\lambda}$ has been used for abbreviation. Several implications follow from this. The quadratic terms both in the numerator (which includes the retardation effects) and in the denominator (which describes the retardation of the depolarization field [121]) will shift the resonance. As stated above, for typical metals well-described by the Drude model, this shift is towards lower energies for larger particle sizes. Interestingly, as Maier notes [116], this also implies that the effect of interband transitions (which are deviations from the Drude model) become less important for larger particles, as the plasmon resonance moves away from that regime. The imaginary term in the denominator finally accounts for radiation damping (direct radiative emission of energy), which is the main reason for larger oscillators to have a weakened strength of the dipole resonance and thus a broadened scattering spectrum [194].

Going back to the quasi-static examinations, the next straightforward step in the expansion of the model considered so far is the investigation of ellipsoids instead of spherical particles, which are a closer approximation to the antenna arms examined experimentally. If the ellipsoid is prolate, i.e. shaped like a rugby ball, two semiaxes b and c are equal to each other and smaller than the third principal axis a. Again the resonance condition can be found via the polarizability, which can be shown [21] to be

$$\alpha(\omega) = \frac{4\pi\epsilon_0 abc}{3} \frac{\epsilon_{\text{ellipsoid}}(\omega) - \epsilon_{\text{med}}}{\epsilon_{\text{ellipsoid}}(\omega) + L_i(\epsilon_{\text{ellipsoid}}(\omega) - \epsilon_{\text{med}})}. \tag{2.53}$$

L_i ($i = a,b,c$) is a geometrical factor, which for the long axis of a prolate ellipsoid can be calculated via

$$L_a = \frac{1-g^2}{g^2}\left(-1 + \frac{1}{2g}\ln\frac{1+g}{1-g}\right), \tag{2.54}$$

where

$$g^2 = 1 - \frac{b^2}{a^2}. \tag{2.55}$$

For the geometrical factors, the sum rule $L_a + L_b + L_c = 1$ holds, so the geometrical factor for the smaller semiaxes b and c can easily be calculated:

$$L_b = L_c = \frac{1}{2}(1 - L_a). \tag{2.56}$$

Again, these approximate tendencies are only valid for very small particles. For such particles, it is thus the aspect ratio of long and short axes that mainly determines the resonance. As above, for larger particles, this can't be the entire truth, as again the resonance would be independent of total size. A correction similar to the one presented for spherical particles above for larger, non-spherical particles of arbitrary shape (which thus again crosses the borders of the quasi-static Rayleigh approximation) has been proposed by Kuwata et al. [102]. Starting with the results of the analytical results for spheres shown in Equation (2.52), they find an empirical equation for the polarizability for arbitrarily shaped particles. Most notably, they find that their fitting parameters to the equation seem to be independent whether they use silver or gold as a material. They reason that this is due to the fact that for the comparingly large particles they examine (approximately 100 nm in size), scattering is dominant and thus material properties become less influential.

Summarizing, it has been shown in this section that Lorenz-Mie-Debye theory can be used to describe the resonances of spherical particles (which are often also termed localized surface plasmon resonances), and that quasi-static models can be used to generate first ideas as to in which regimes optical resonances for small particles of spherical and ellipsoidal shape can be expected. In the next section, particle resonance as the solution of a waveguiding problem shall be examined, which is an especially interesting approach as it can be used to bridge the gap between classical antenna theory and optical antennas.

2.2.3 Antenna Resonance as a Waveguide Problem

In the following, an infinitely extended cylindrical wire of radius R is considered as a waveguide. Its fundamental propagating mode is rotationally symmetric and usually termed TM_0 mode. Such an infinitely extended wire does not radiate due to its translational symmetry. If, however, a short piece of wire of length L is cut out, it can emit electromagnetic radiation and thus becomes an antenna.

The wave propagation within the short piece of wire can now be used to come up with a resonance condition. Basically, the short wire forms a cavity with reflecting ends (the wire facets), such that constructive interference of the reflected waves can occur. The shortest length at which such a constructive interference occurs (i.e. the round trip phase gain after twice the wire length is equal to 2π) is thus the fundamental scattering resonance frequency of the short piece of wire. The only other consideration that needs to be made is that upon reflection at the facets, the light may and will undergo some phase shift ϕ_{ref}. Mathematically, the resonant wire length L_{res} for the fundamental Fabry-Pérot resonance can be expressed as [18]

$$L_{\mathrm{res}} = \frac{\pi - \phi_{\mathrm{ref}}}{\beta}, \tag{2.57}$$

where the mode wavelength $\lambda = \frac{2\pi}{\beta}$ is calculated from the propagation constant β of the wave.

For a linear optical antenna modeled as a thin cylindrical wire of radius R that is significantly smaller than the wavelength, and a dielectric function described by a free-electron gas according to the Drude model (which is a reasonable assumption as long as interband transitions don't occur, as pointed out in previous sections), an analytical solution (under some approximations) for the resonance has been found [137]. It reads as

$$2L_{\mathrm{res}} = 2\pi R\left(a_1 + a_2\frac{\lambda}{\lambda_p}\right) - 4R, \tag{2.58}$$

where λ_p corresponds to the plasma frequency of the material and a_1 and a_2 are material specific parameters that depend only on the infinite frequency limit of the dielectric function of the material and the dielectric constant of the surrounding medium. The offset of $-4R$ is an approximate inclusion of the phase change, which can also be considered an apparent increase in antenna length due to the reactance of its ends. With all parameters introduced for numerical evaluation, the resonant arm length for an incident wave of frequency ω can be calculated as

$$L_{\mathrm{res}} = \pi R\left[13.74 - \frac{2}{\pi} - \frac{0.12}{\epsilon_{\mathrm{med}}}(\epsilon_\infty + 141.04\epsilon_{\mathrm{med}}) + \frac{0.12\hbar\omega_{\mathrm{p}}\sqrt{\epsilon_\infty + 141.04\epsilon_{\mathrm{med}}}}{\epsilon_{\mathrm{med}}}\frac{1}{\hbar\omega}\right], \tag{2.59}$$

where as before R is the rod radius, ϵ_{med} is the refractive index of the surrounding medium, ϵ_∞ is the infinite frequency limit of the dielectric function, and ω_{p} is the bulk plasma frequency. Strictly speaking, the scaling is linear only with wavelength (and thus with inverse frequency), but for investigations in the optical regime only, the scaling is almost linear with frequency as well.

Such resonances have been measured experimentally for gold rods, and the scaling law as given by the model described above has been confirmed by Dorfmüller et al. in the near-infrared regime [49]. For aluminum, the solution is improved by using measured values of the dielectric function, as there are significant deviations from the Drude model solution around 800 nm (see e.g. Figure 2 in reference [137]). One of the most interesting conclusions of Novotny's analytical approximation is that it can be expanded to also consider a perfectly conducting cylinder. In the limit of vanishing R (the cylinder becomes a line), the expression derived then converges towards $2L_{res} = \lambda$, which is exactly what is obtained from classical antenna theory, as had been described in Chapter 2.1.2. The waveguide analysis is thus a tool that bridges optical antenna behaviour and classical antenna theory and also shows that the plasmonic resonance and classical antenna resonance converge for single wire antennas if the main additional assumptions for optical frequencies (material specific damping instead of perfect conductivity and consideration of finite thickness instead of pure surface currents) are dropped.

A very interesting analysis of what happens if the linear thin wire antenna geometry is adapted for plasmonic resonances (i.e. if volume currents instead of surface currents are considered) has recently been presented by Dorfmüller et al. [48]. Using an approach similar to the classical textbook analysis of linear dipole RF antennas where now a field inside the antenna structure is allowed, they obtain an expression for current and phase of a single-arm optical antenna, which they confirm with experimental data. The model can even be used to predict the radiation patterns of such structures (see also the supplementary information of reference [48]) and is expecially appealing as it should be adaptable for the transition to the RF regime.

2.2.4 Two-Arm Antenna Resonances

So far, single-arm optical antennas have been considered in this theoretical overview. It is now time to add the last antenna element used in the later experiments to the analysis: the antenna gap. For classical antennas, we have seen earlier that a gap is typically only a very small perturbation for an antenna. However, this is mostly due to the fact that the gap is very small compared to the wavelength and generally the antenna dimension, which is not the case for typical optical antennas due to fabricational issues. A generally achievable gap size of about 20 nm as mostly used in the experiments presented later constitutes more than 10% of the total antenna length of structures operating in the visible regime for gold and aluminum nanoantennas. Another way of looking at this situation is impedance matching: For classical antennas, the gap perturbation is eliminated by the matched impedance of the feed lines towards the antenna arms, such that the antenna fundamental resonance is found for two segments of length $\frac{\lambda}{4}$ each. For nanoantennas, however, the gap is not impedance matched, such that, while antenna coupling still happens, the resonance shift is much different [137].

As two structures couple to produce a new resonance (in fact we will see in a few lines that actually two new resonances are formed), it seems logical to approach the problem from a point of view of two oscillators approaching each other. They may either be oscillating in phase (which is the situation just described), but might also oscillate out of phase (which can typically not be observed, as both antennas are driven by the same electromagnetic wave at the same phase). As the second behaviour is conceptually valid and can even be experimentally observed [83], the concept of *plasmon hybridization* [136] has been accepted as a valid description for two-arm antenna coupling, even for coupling of tilted structures [66]. Conceived as an electromagnetic analog of molecular orbital theory [153], the model successfully points out the resonance behaviour of two-arm structures of the so-called bonding (red-shifted) and anti-bonding (blue-shifted) modes. The splitting behaviour becomes more prominent when the gap size is decreased. The anti-bonding modes are often also called *dark modes*, as they cannot be excited using symmetric excitation patterns, as the two rods oscillate out of phase in that case, compared to in-phase oscillations of the bonding *bright mode*. It will thus be spectral red-shifts that are expected to be observed in linear scattering experiments for two-arm antennas coupled by a small gap. Simulations also show that the resonance broadens for the bonding modes of two-arm antennas compared to the single arm counterparts, even more so for decreasing gap sizes [18]. This also implies that the quality factor of Q of such a resonance, defined as $Q = \frac{\omega_{\text{res}}}{\Delta\omega}$, where $\Delta\omega$ corresponds to the full width at half-maximum (FWHM), is lower for two-arm structures. Thus two-arm antennas are a resonator of lesser quality, which on the other hand means they have increased losses, either desired (radiative) or undesired (Ohmic).

Generally, the resonance shifts are fairly small for realistically achievable gap lengths, so the main resonance is still determined by the resonance of the single-arm structure, with a perturbation induced by the antenna gap as described above. In the spirit of the Thévenin equivalent circle of classical RF antennas, it should also be mentioned that the inclusion of material near the antenna, especially in the antenna gap, can be looked upon as a change in antenna (input) impedance [56, 5], and will thus also create a shift in the antenna emission spectrum similar to antenna length and gap size influences.

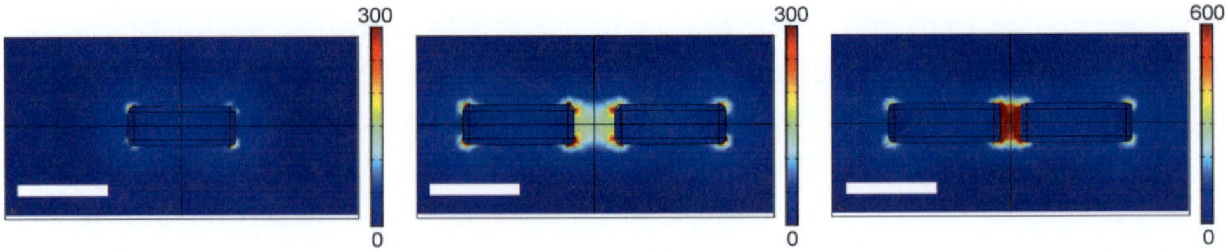

Figure 2.10: Near-field intensity enhancement (given in false colours, note the different scale for the third structure) for 55 nm arm length single-arm and two-arm antennas, the latter two with a 20 nm and 10 nm gap. Scale bar length is 50 nm. (Simulations courtesy of Carola Moosmann)

Near-field Behaviour at the Antenna Gap

So far, antenna far-field resonances have been considered exclusively. To make optical antennas useful sensors, their near-field resonance behaviour and localization in conjunction with locally enhanced fields plays a major role. Analytical models can provide such information as well, but the fastest and easiest way to obtain near-field data for single nanostructures is the straightforward implementation and solution of Maxwell's equations, respective material properties and geometries in a numerical simulation tool. Numerous such simulations have been performed [2, 71, 148, 18], and the findings can be quickly summarized: The near-field resonance occurs at (almost) the same frequency as the far-field resonance, and the strongest enhanced fields are found around singularities of the structure, i.e. at its ends as well as (for two-arm structures) in the antenna gap. In fact, the fields in a two-arm antenna gap are significantly higher than near simple ends of single-arm structures. Figure 2.10 gives a comparison of the near-field intensity (electric field squared in the center-plane of the antennas) for a 55 nm arm length gold single-arm antenna with two-arm antennas with a 20 nm and 10 nm gap, respectively. All antennas are excited at their specific resonant frequency. The extremely enhanced fields for the gap antennas are clearly visible. Note the different intensity scale for the 10 nm gap antenna.

The increase in resonance shift as described in the previous section for two-arm antennas with decreasing gap size thus goes hand in hand with a shift of the near-field localization. In fact, an intensity enhancement of about 10,000 can be expected in a 4 nm gap of a resonantly driven 100 nm arm length antenna [17]. The coupling of two antenna arms thus provides a significant advantage in field concentration and can serve as an efficient design tool in antenna fabrication, which is why fundamental properties of two-arm antennas are extensively investigated in this work. Effects of the gap on the scattering response are observed in Chapter 6, while near-field investigation results are presented in Chapter 8.

2.3 Describing Optical Antennas

Now that both radio frequency approaches as well as the localized-plasmon based description of nanoparticle resonances have been extensively introduced, it is time to reconsider the questions asked at the beginning of this chapter. Which framework is suited best to describe optical antennas?

A simple localized surface plasmon model based on quasi-static descriptions is probably best-suited to generate ideas as to which materials can realistically be considered for antennas in the optical regime. Coarse estimates for resonance behaviour, scattering, and absorptive losses can be generated with relative ease.

Classical antenna theory also falls short on the quantitative estimation of antenna resonances, but offers an extensive framework to especially work with geometries more complicated than the single wire case presented in this work. Generally, the terms describing antenna behaviour such as radiation pattern, antenna resistance, and gain, keep their meaning in the optical case, offering valuable insights and easy description, and especially a well-developed language to speak of antenna effects. Efforts to also transfer equivalent circuit language to optical antennas are also valuable [57].

In my opinion, the best way to look at optical antenna behaviour are hybrid approaches considering measured and/or modeled dielectric functions of metals and working them into classical waveguide and antenna theory. The frameworks presented by Novotny [137] and Dorfmüller et al. [48] hold great promise for future antenna design, as they can be easily adapted for different materials, and, especially the theory in reference [48], should also be adaptable to more complicated geometries, yielding additional insight into optical antenna radiation patterns and higher order optical resonances.

Until these frameworks are completely established, numerical simulations based on Maxwell's equations are a time-intensive, but nonetheless valuable and indispensable guideline for optical antenna design. Most important, however, remain experiments performed with prototype nanostructures, which shall be presented in the following chapters. Before those will be considered, nonlinear effects not described by any of the approaches presented so far need to be introduced, as they form a substantial part of the analyses performed.

2.4 Nonlinear Optical Effects in Optical Antennas

So far, we have considered antennas as materials linearly responding to photons, absorbing or scattering them, thereby possibly giving rise to enhanced electromagnetic field intensities very near to the antenna surface. However, the nonlinear properties of the metals typically used for optical antennas can also come into play and lead to effects substantially differing from what can be explained by the theory outlined previously. Nonlinear plasmonics, as one might term the study of such effects [15], is a recently emerging field striving to exploit this behaviour.

There are two main types of nonlinear effects that are currently investigated. The first effect can be explained in the framework of nonlinear susceptibilities and includes all types of frequency conversion effects, which for the most simple case is second-harmonic generation. As is extensively explained in the excellent book by R. W. Boyd [28], the polarization as described for a purely linear material in Equation (2.31), is a simplification only true for comparably low excitation powers. To be more exact, one would need to write

$$P(t) = \epsilon_0[\chi^{(1)}E(t) + \chi^{(2)}E^2(t) + \chi^{(3)}E^3(t) + \cdots], \tag{2.60}$$

where $E(t) = E_0 \exp(-j\omega t) + E_0^* \exp(j\omega t)$ and the respective $\chi^{(i)}$ are the (generally tensorial) susceptibilities for the higher order effects, which give rise to mixing of frequencies and generation of higher harmonics. For simplification only an instantaneous material response has been assumed (which makes the medium lossless and dispersionless). The respective χ coefficients of higher order are very small, but for appreciable input powers, they may become important and even dominate the response. Symmetric materials require some terms to be 0, but generally speaking, some non-linearity is always available. Gold has a very high third order nonlinear susceptibility [141] on the order of 0.2 nm^2/V^2 (which is more than two-orders of magnitude higher than the one of $LiNbO_3$) [154], which enables higher order frequency conversion schemes. Indeed, enhanced second-harmonic generation (that is, generation of light with twice the energy than the incident photons) has been observed for roughened gold surfaces [196] and at the extremities of gold tips [24], gold third-harmonic generation has been shown [113], and four-wave mixing has been demonstrated for coupled gold nanoparticles [45]. These effects have the advantage of allowing an easy separation of excitation and emission light by spectral filtering of the excitation source with e.g. a band-pass filter only allowing the second-harmonic light to pass through. Some second-harmonic emission from gold was also observed during this thesis, the effects are described in Chapter 9.

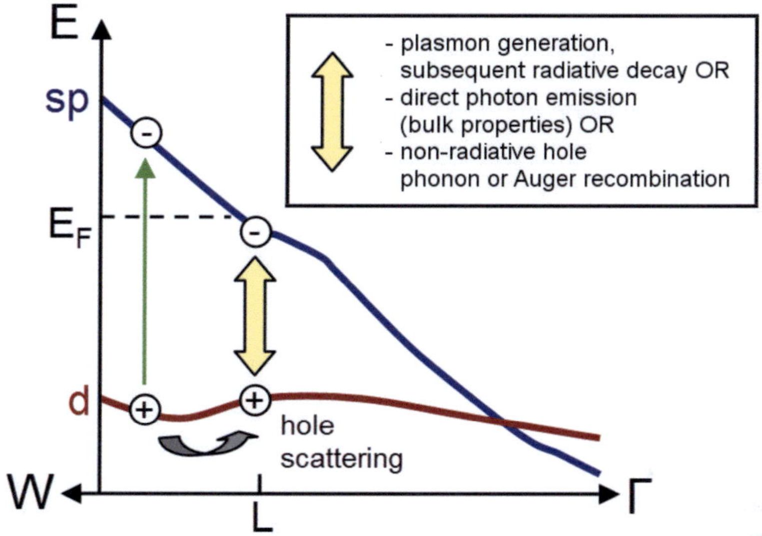

Figure 2.11: Hole generation through photoexcitation (green arrow) and subsequent hole scattering (grey arrow) gives the possibility of recombination of electrons below the Fermi level with that hole. The recombination (light blue arrow) may occur by transferring the energy to a particle plasmon, which may then decay radiatively, by direct photon emission, or the hole may just undergo Auger or phonon processes and decay non-radiatively. (Image inspired by [51, 27])

A second type of nonlinear effect is the emission of broadband light under high intensity laser illumination, where the response is distributed over a wide range of frequencies. In addition to higher harmonics, Mooradian discovered in 1968 [127] that also a broadband response is created by bulk gold and copper under single-photon excitation, G. T. Boyd et al. extended this to multi-photon processes in bulk gold, silver, and copper in 1986 [27] and also expanded the theoretical explanation. For gold and silver, they were able to explain the spectral shape of the emission by the band structure of bulk gold. For some spectra obtained using multiphoton excitation, especially for silver, they also assumed that a contribution of volume and surface plasmons is responsible for parts of the spectral response. In general, their explanation involves the generation of electron-hole pairs, which then relax according to the band structure: the electrons recombine below the Fermi level with holes in the d bands around the X and L symmetry points.

Nanostructuring capabilities have strengthened the interest in the original bulk and surface results in recent years, searching for broadband light generation in nanostructures. Several types of effects have been observed, most notably emission for nanostructures very similar to the bulk results from single rods [84], higher-order nonlinear effects [128] and emission shaped by plasmon resonances for single nanoparticles [51, 23]. Figure 2.11 gives a detailed overview of the types of nonlinear relaxation expected around the L symmetry point for gold as suggested by Dulkeith et al., considering both bulk structure properties as well as plasmon generation. Chapter 7 will show experiments to gain further insight into such processes, especially for two-arm antennas.

2.5 Other Optical Antenna Geometries

In general, this thesis will only deal with linear single-arm and two-arm antennas and combinations thereof. However, there are several other types of optical antennas that have mainly been derived from their RF counterparts that should briefly be mentioned.

One antenna type that has gathered much interest is the so called bow-tie antenna, two triangles pointing at each other. While it may look easier to fabricate small antenna edges using such structures, in the ultimate fabrication limit the width of the antenna gap will be the same as for two-arm antennas. Bow-tie structures are more broadband, which can give advantages for some applications [162]. A bow-tie antenna fabricated using focused ion-beam milling on an atomic force microscopy tip [58] is shown in Figure 2.12 a).

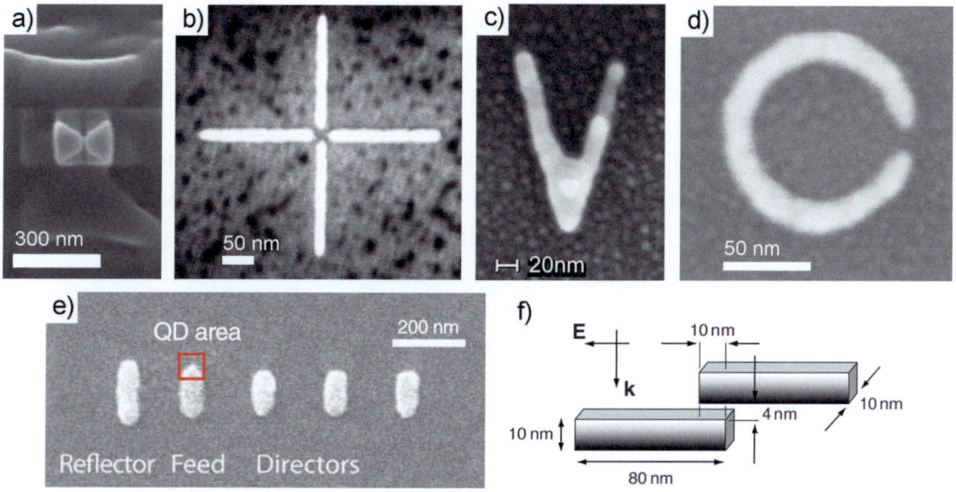

Figure 2.12: SEM images (except f)) of various optical antenna geometries: a) bow-tie antenna (reprinted from [58], with permission from IOP Publishing and the corresponding author, DOI), b) cross-antenna, c) v-antenna (reprinted from [94], with permission from Elsevier), d) split-ring antenna, e) Yagi-Uda antenna (from [42], reprinted with permission from AAAS), f) stacked optical antenna (reprinted with permission from [151], copyright 2011, American Institute of Physics).

Recently, cross-antennas that are to be used for circularly polarized light have found some interest [142, 19]. An example is given in Figure 2.12 b). Fabricational issues (as shall be seen later) arise from the cross-exposure within the gap area of such strutcures. In the context of this thesis, they will only be presented as devices which can be used to independently tune frequencies for two polarization components for linearly polarized light.

V-shaped [94] and split ring antennas (Figure 2.12 c) and 2.12 d)) have also been given some attention; especially split rings are however more often found in metamaterials applications and their single entity behaviour has only rarely been investigated, even though they should also generate interesting localization properties, again mostly for circularly polarized light. For that purpose, nanoscale spirals have also been considered recently [199].

Yagi-Uda antennas (2.12 e)) are a very common concept in RF antenna design, as they have a favorable unidirectional emission pattern compared to other structures. This behaviour can also be reproduced for optical antennas, as has been shown very recently both theoretically [82] and experimentally [98, 50]. Such devices have also been combined with single emitters to couple light from the near-field to the far-field [42, 41].

Pohl and Novotny have proposed a stacked optical antenna [151] as depicted in Figure 2.12 f), where two antenna arms are fabricated in two layers on top of each other, which could also generate enhanced near-fields in the gap between the two structures.

With all these antenna types available, why are mainly two-arm optical antennas investigated in this work? The answer to this question encompasses the many properties of antennas and the field of plasmonics described earlier. Compared to all other geometries, they are the fundamental prototype for antenna investigations as they make available efficient impedance matching between optical far-fields and near-fields for selected frequencies (possibly in combination with chromophores), they give strongly enhanced field concentration in the antenna gap, and also have this field localized in a volume with a size significantly below the diffraction limit. All other geometries, where some examples have been listed above, may yield advantages in one of these fields (often with drawbacks in another), but generally the two-arm geometry enables the most fundamental study of optical antenna properties and already brings with it a tremendous amount of applications. Some general fields in which the use of optical antennas will be beneficial are described in the following section.

2.6 Applications of Optical Antennas

While this thesis gives experimental results on very fundamental properties of optical antennas, all such properties are to be used at some point for functional devices. It is thus important to single out the most likely fields of application for optical antennas.

Biological sensors [68] are one of the most promising fields. With electromagnetic near-fields highly localized in nanogaps as outlined above, different types of sensor concepts can easily be imagined. For spectroscopic applications, apertureless scattering near-field optical microscopy probes [140, 8] can be enhanced by optical antennas, changing emission intensity and lifetime [20, 59]. The hot spot in two-arm antennas as investigated in this work can be used to detect molecular interactions slightly changing the refractive index [68]. Following the same path, enhanced Raman scattering from such enhanced near-fields can yield additional molecular information from very small volumes [197, 171, 167], as exemplified in Figure 2.13 a). The key in all such endeavours is to find the right distance between antenna and fluorescent element to be observed to enable enhanced emission and avoid quenching effects [16].

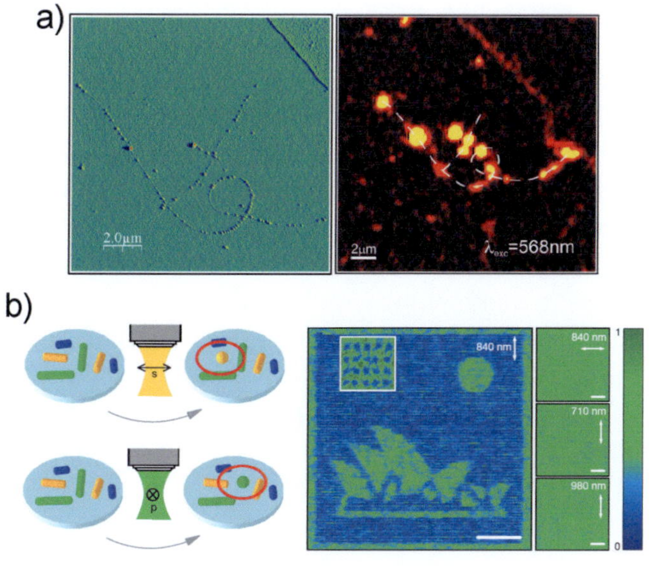

Figure 2.13: Two optical antenna applications: a) enhanced Raman scattering; left: single wire carbon nanotubes are covered with metallic nanoparticles (AFM image), right: at the position of the particles, the photoluminescence signal and the Raman signal are enhanced (reprinted with permission from [167], copyright 2008, American Chemical Society), b) photothermal data storage; left: single-arm antennas are deformed with wavelength and polarization selectivity, right: luminescence read-out is only possible at the chosen wavelength and polarization (reprinted by permission from Macmillan Publishers Ltd: Nature, [200], copyright 2009).

Optical antennas are also thought of to be a formidable tool in enhancing the efficiency of photovoltaic cells [18]. The key idea here is to use the greatly enlarged cross sections of antennas at resonance to capture more incident light into the active layers. At the current stage of research, relatively large metallic particles are mostly used to increase optical path lengths in the active material, thereby also increasing the probabilty of conversion into energy [33]. A different approach would be to directly convert energy incident on optical antennas into a DC current by the use of high-speed quantum tunneling diodes. Such rectenna approaches have found great interest about 30 years ago in the infrared regime [157], and are currently investigated again for the visible [14, 184].

Optical trapping using the enhanced near-fields of optical antennas has also been proposed [138], basically applying them as optical tweezers. For example, trapping of living bacteria and other particles near gold disks has been demonstrated [155].

The exploitation of heating effects in resonantly excited metallic particles could also find uses in therapeutical applications. Particles selectively bonded to tumor cells are heated to destroy the latter as photothermal therapy [105].

Finally, densely packed data storage has been shown exploiting plasmon resonances of single rods to selectively melt and thereby change the shape of the nanostructures [200]. The principle is visualized in Figure 2.13 b). Of course, these are only a few examples for the still daily growing number of applications.

Chapter 3

Luminescence and Semiconductor Quantum Dots

Very small nanostructures able to emit a sequence of single photons can, as already hinted in the previous chapter, be used as nanoprobes to give specific information about local electromagnetic field strengths. This is possible since they will only respond to significant electromagnetic intensity present at their very local position. Thus if the local excitation intensity changes, their emission intensity will as well. Beside their use as field sensors, they can also be thought of as single-photon sources: If only one photon is emitted at a time, a stream of distinct quanta is sent out, which can be used for applications such as quantum cryptography.

Single atoms are such single photon emitters, but they are typically too small to handle and have insufficient fluorescence efficiency due to a relatively small absorption cross section. An ideal quantum single photon emitting probe would thus be some kind of artificial atomic system with high efficiency.

So-called *quantum dots* made from semiconductors can be used for that purpose. Most fundamentally, they are nanoentities artifically engineered to emulate atomic two-level systems, where excitation light can be used in an absorption process to promote electrons to the higher energetic level, from which they can decay radiatively, emitting a single photon. Typically, such quantum dots follow an ordinary fluorescence exciation scheme. Understanding luminescence processes is also important for the excitation of gold nanostructures using a pulsed laser performed in later chapters, which is another reason why they will be introduced in the following.

3.1 Fluorescence and Phosphorescence

The excitation and decay of excited electronic states is conveniently depicted in a *Jablonski diagram*, as seen in Figure 3.1.

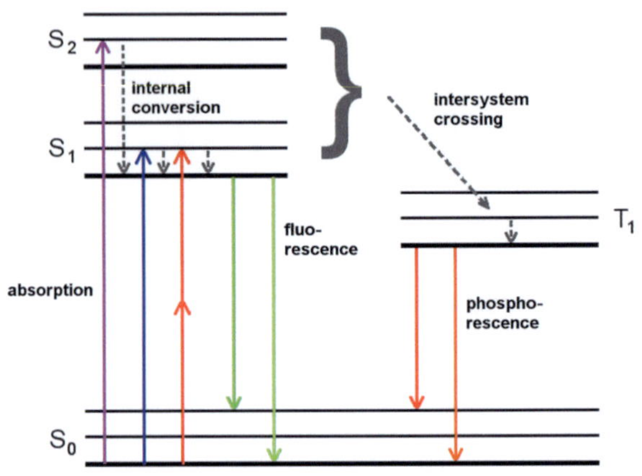

Figure 3.1: A simple Jablonski diagram containing absorption, fluorescence, and phosphorescence. Vibrational relaxations are depicted by short dashed arrows. (Image inspired by Lakowicz [103])

Obviously, the transitions in such an artificial system are not quite as simple as introduced above. Both the ground state (the lower level) and the excited state (the upper level) also exhibit vibronic side bands. From those vibronic levels, the electrons will decay non-radiatively to their corresponding lowest level. This is why relaxation (and thus radiative emission) always yields higher wavelength (lower energy) emission, an effect which is also known as Stokes-shift: Some energy is lost after absorption and basically transferred into heat. The radiation obtained from the relaxation is called *fluorescence*. Of course, some electrons will find other non-radiative pathways to give up their energy, so it is useful to define a *quantum efficiency* η for fluorescence processes [139]:

$$\eta = \frac{k_\mathrm{r}}{k_\mathrm{r} + k_\mathrm{nr}}. \tag{3.1}$$

k_r and k_nr denote the radiative and non-radiative decay rate, respectively. The inverse of the decay rate is commonly known as lifetime, so e.g. the radiative lifetime (sometimes also called natural lifetime) t_r is defined as $t_\mathrm{r} = \frac{1}{k_\mathrm{r}}$. The higher η, the more useful is the fluorescent object as a sensor, as it will convert more inciding light into a measurable response. It should be noted that when a decay is observed, the measured lifetime will be statistically distributed like an exponential function, of which the so-called *apparent lifetime* $\tau = \frac{1}{k_r + k_{nr}}$ is the inverse decay constant. It is important to realize that the apparent lifetime as observed in an experiment will thus be influenced both by changes in natural lifetime and the availability of non-radiative decay channels.

Sometimes, the already excited electron can also undergo a flip in spin and energy level (using some additional vibrational energy, which is why this process is called *intersystem crossing*). It is then said to be in a triplet state, from which it will take longer to relax, as the transition back to the ground level is spin-forbidden. This is the process known as *phosphorescence*.

Normally, if the energy of the incident photon is too low to promote the electron to the higher level, no absorption will occur. However, several photons may combine to yield the energy required to excite an electron, as exemplified by the two combined red absorption energy arrows. When the energy of more than one photon is used up, this process is called *multiphoton absorption*. The special case with two photons as indicated by the two arrows in Figure 3.1 is called *two-photon absorption*. For the process to happen, the energy from two quanta has to be provided at almost the same time, so the probabilities for such a process to happen are only significant for relatively large incident powers. The order of the absorption process can for example be derived from the dependency of the emitted power to the incident power: For a two-photon process the emission power depends on the square of the incident intensity [103]. Such two-photon processes will be observed and analyzed for resonantly driven metallic structures in Chapter 7 and for non-resonant gold structures in Chapter 9, albeit the emission processes will be more difficult to describe than what has been shown here in the simple Jablonski diagram.

3.2 Quantum Dot Functionality

Now that it has been shown how different available energy states determine fluorescence (and phosphorescence) properties, it is time to explain how they can be engineered in quantum dots and why such structures are advantageous over many other molecules that show such properties. The latter is easily answered – they can be made to be very stable even under high excitation powers and are much less likely to degrade or bleach when used continuously, and their emission spectrum can be tuned very finely during the fabrication steps. Taking a look at the working principle of quantum dots will explain why.

Quantum dots are typically fabricated using semiconductors. As is commonly known [36], these materials have a so-called band-gap of forbidden states, i.e. there are only electronic states allowed above and below a certain level. In between, there are no allowed states. For such materials, states in the lower bands are mostly occupied, while only few states in the upper bands are. (E.g. optical) excitation can promote an additional electron from a lower band and state into a higher band. From there, it may (among other things that may happen non-radiatively, such as Auger recombination) relax radiatively, just like seen before, emitting a red-shifted photon. The band-gap determines the emission wavelength.

A very small sphere of semiconductor material will suddenly show a somewhat enlarged transition, and the shift goes even more towards the blue for smaller entities. What triggers this unexpected behaviour? Of course, this problem can be solved quantitatively (see e.g. [158]), but at this point a qualitative approach as presented e.g. in [131, 4, 52, 168] shall suffice. If one considers the small sphere as a box, then the particles inside, namely electrons and holes, become more and more fixed in terms of their position. By Heisenberg's uncertainty principle, such a particle confined in space (thus the term *quantum confinement effect* for such behaviour) needs to have additional momentum and thus kinetic energy, leading to an increase in energy gap size.

So what exactly is *very small*? It all depends on the spatial separation of electrons and holes in generated electron-hole pairs (termed excitons). Their separation is also called an effective Bohr radius, as it mimics the behaviour of states in a simple hydrogen atom. This exciton Bohr radius (or diameter) can be calculated for several semiconductors and is dependent on effective masses of electrons and holes. To give an example, for CdSe one obtains a Bohr radius of 5.6 nm [55]. As soon as the size of the particle is on the size of or smaller than that radius, the particle in a box effect becomes dominant and the gap size is already significantly increased.

Under some approximations, Brus has developed a simple formula for band gap estimation ([30], see also e.g. in [131])

$$E_{\mathrm{g,QD}} = E_{\mathrm{g,bulk}} + \frac{h^2}{8R^2}\left(\frac{1}{m_{\mathrm{eff,e}}} + \frac{1}{m_{\mathrm{eff,h}}}\right) - 1.8\frac{e^2}{4\pi\epsilon_0\epsilon R}, \tag{3.2}$$

where R is the radius of the semiconductor particles, while $m_{\mathrm{eff,e}}$ and $m_{\mathrm{eff,h}}$ are the respective effective masses. The middle term is the one described by the particle in a box analogon above, while the last term introduces corrections for Coulomb-interaction effects, which can be neglected for very small particles as defined above. In summary, quantum dots of a few nanometers in size can thus mimick the absorption and emission behaviour of atoms, which is why they are sometimes called *artificial atoms*. On a side note, these quantum-confinement effects are typically not observable for metals as discussed in the previous chapter, as the bands are continuous in metallic particles even for relatively small sizes. However, once the conduction electrons are confined below the Fermi wavelength of an electron (smaller than 0.7 nm), discrete electronic transitions will occur, as observed e.g. for small nanoclusters consisting of five up to 31 gold atoms [198].

Quantum dots also show *blinking* behaviour. Between time-frames in which they efficiently absorb and emit light, there are time intervals where no light is emitted and the dots appear dark. Typically, this fluorescence intermittency is explained by net charge fluctuations present inside or around the nanocrystal [135, 64]: If an excited electron or hole moves into trap states, there are delocalized charges in the nanocrystal core. Subsequently excited electron-hole pairs now have additional non-radiative Auger process-like decay paths available, which are much faster [95] than the usual radiative decay. When the trap state is left, the quantum dot recovers to its previous radiation abilities.

3.3 Semiconductor Quantum Dot Fabrication

The quantum dots used in this thesis were purchased, so only a very brief introduction into their fabrication will be given and the interested reader is lead to other publications for more information [108, 149, 43, 139].

Single quantum dots are often generated using epitaxial self-growth: Nanofacets and nanoislands form due to lattice-mismatch that induces mechanical strain which favors specific configurations of material [107]. As an example for such a growth process (there are numerous other possibilities), when GaAs is deposited on InAs, which has an about 7% larger lattice constant, at first a thin layer will form. But after a few layers, suddenly some small islands will appear, which, if growth speed and timespan is chosen accordingly, can have quantum dot properties. This growth scheme for quantum dots is known as Stranski-Krastanov method [139].

The second important fabrication scheme is chemical synthesis. This bottom-up approach has the advantage that great numbers of quantum dots can be prepared in a colloidal solution, which is also the process applied for the quantum dots used in this work. High-temperature synthesis in coordinating solvents is the method of choice in many cases [132]. Defects and impurities can introduce additional unfavorable states within the desired band-gap of the quantum dot. As these can most easily occur at the surface of the structure, they are best avoided by an additional covering shell layer. The quantum dots used throughout this work have a CdSe core and an outer shell of a ZnSCdS

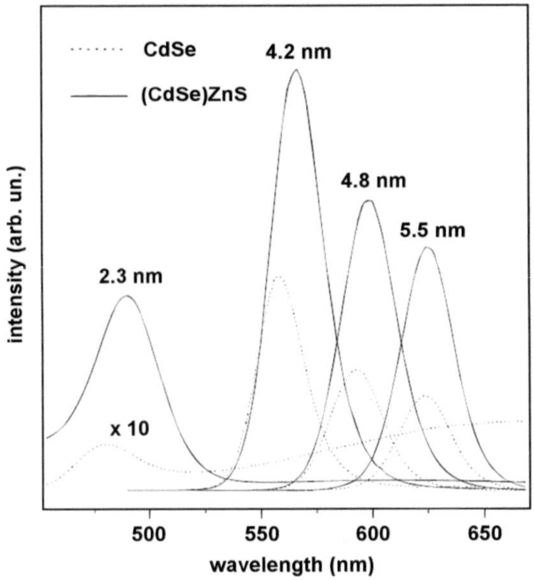

Figure 3.2: CdSe quantum dot photoluminescence spectra for various core sizes with and without a ZnS shell. The coating strongly increases the photoluminescence and only introduces a small shift in peak wavelength. The CdSe core size is given above the respective spectra. (Reprinted and adapted with permission from Dabbousi et al. [43], copyright 1997, American Chemical Society.)

alloy. The shell material has a larger band-gap than CdSe, thereby additionally confining the generated excitons to the core, leading to further surface recombination minimization [168]. An example for the enhancement in emission yield is depicted in Figure 3.2. For larger core sizes, the photoluminescence is shifted to the red. While some shift in peak wavelength is also caused by the coating, its main effect is the strongly enhanced emission signal. Structures with this setup are also called type I quantum dots. In contrast, type II quantum dots have valence as well as conduction bands in the core lower (or higher) than in the shell (for more details see e.g. reference [91]). Compared to many other luminescent substances, quantum dots are very stable under exposure and can be used over very long timespans (even weeks or months), if properly covered. Thus photobleaching is not an issue in many experiments for semiconductor quantum dots, even though they may e.g. show photobluing, that is a spectral blueshift of the emitted light, under longtime exposure [178].

3.4 Some Applications of Quantum Dots

Due to their relatively easy fabrication and their favorable optical properties, semiconductor quantum dots have found numerous applications. Besides the obvious use as slowly photobleaching labels in biomedical investigations [168], one of the most promising fields is quantum dot lasing. Quantum dots offer a relatively sharp Gaussian-like spectral lineshape and a large tunability during fabrication. Additionally, low threshold current densities can be reached, with the lowest threshold quantum dot laser diodes obtained from Stranski-Krastanov self-organized growth [31].

Enhanced solar cells using quantum dots have also been proposed. The main idea here is that a one-material semiconductor can still show multiple band gaps from different dot sizes which can be combined to increase absorption efficiency from the solar spectrum [75]. Quantum dot light emitting devices have been suggested as well, which are advantageous due to the combination of relatively pure colours offered by quantum dots and their solution processibility [38, 6]. Also, the phosphors often used to achieve white-light emission from ordinary LEDs can be replaced by layers of quantum dots, as e.g. sold by QD Vision, Inc.[1] Finally, a prototype full-colour quantum dot display has been presented recently [92].

Quantum dot infrared photodetectors [31] and applications for quantum optical computing due to the emission of single photons [114] are also foreseeable. In this thesis, quantum dots are used to map out the two-photon point-spread function of the confocal characterization system available. Additionally, it is shown that single quantum dots under two-photon excitation also show antibunching behaviour. This makes them suitable candidates even at those excitation conditions to be used in conjunction with optical antennas, which may enhance their emission properties. Also, results on the combination of such quantum dots with non-resonant rod structures as sensors for enhanced electromagnetic fields will be presented in Chapter 9.

[1] http://www.qdvision.com

Chapter 4

Antenna Fabrication and Topographical Characterization

After the description of the theoretical background and the general behaviour of nanoantennas in Chapter 2, it is now time to present the real structures used throughout this work and how they were obtained. Even with today's daily increasing technological possibilities, it is still challenging to reproducibly fabricate structures with dimensions of only a few tens of nanometers. Only a very limited number of technologies is available for such tasks. While thin layers can easily be fabricated using technologies such as spincoating, doctorblading, chemical vapor deposition, thermal evaporation, or sputtering (to only name a few) [118, 187], the fabrication of structures limited in more than one dimension leaves significantly less choices. Chemical processes allow for the fabrication of many materials with simple shapes (for example spheres or rods), but arbitrarily aligned particles of arbitrary shapes for arbitrary materials are again harder to manufacture. Nonlinear laser processes can allow for structuring of photoresists down to the diffraction limit and even below, but still, one is limited to several tens of nanometers resolution. Thus only focused ion beam milling, nanopatterning via atomic force microscopy tips, and electron beam lithography are commonly used to generate functional prototype nanostructures. The patterning via atomic force microscopy is inherently slow (as the speed of the cantilever tip is limited), so only focused ion beam milling or plain electron beam lithography can be easily used to generate a large amount of structures. Typically, focused ion beam milling can only be used after a coarse structuring via electron beam lithography has been performed, which in total would then require a two step process. Thus, a fabrication protocol only involving electron beam lithography would be highly preferential. For the structures used in this work, a protocol was developed that can be followed to generate structures with minimal feature sizes down to 20 nm using electron beam lithography exclusively, which will be presented in the following. Like for all electron beam lithography schemes, a resist is first applied to a sample, which is then exposed by electrons at specific points. In a development step, the exposed areas are removed (for a so-called positive resist as used in this work; it is also possible to only remove the non-exposed areas using a negative resist). A thin layer of some material is then applied to the sample using some deposition technique, which will come into contact with the sample where the resist had been removed. In a final lift-off step, the deposited material and the remaining resist are stripped off the sample at all other points, leaving structures on the sample.

Two similar, but slightly distinct engineering approaches were developed. While gold antennas were produced on a glass cover slip covered with indium tin oxide (ITO), aluminum antennas were not only fabricated on ITO, but also directly on glass with a conductive mask to allow for alignment of the electron beam lithography system. Thin glass as a transparent substrate was used in all cases as it allows for optical characterization even with oil immersion objective lenses as well as manipulation of the sample from both sides, which is important for many of the experiments conducted later. All electron lithography steps were performed using a Raith e_LiNE system (Figure 4.1), which was made available by courtesy of the Institute of Micro- und Nanoelectronic Systems (Prof. Dr. Michael Siegel, Dr. Konstantin S. Ilin). In the following, the antenna fabrication scheme used in this work is detailed. Alongside, results of the topographical characterization will be shown to demonstrate the high accuracy as well as reproducibility of the fabrication scheme.

4.1 Gold Antenna Engineering on an ITO/Glass Substrate

For reasons already pointed out in the previous sections, gold is a highly attractive material for optical antennas. For nanostructures, it is usually evaporated to allow for small grain sizes and smooth patterns, a process which requires a substrate with a structured resist that it will also adhere to during a lift-off process once the evaporation has been performed. For example, a substrate made from silicon would be ideal for these structures with only these constraints, as its surface is very smooth and gold is easily lattice-matched. However, the

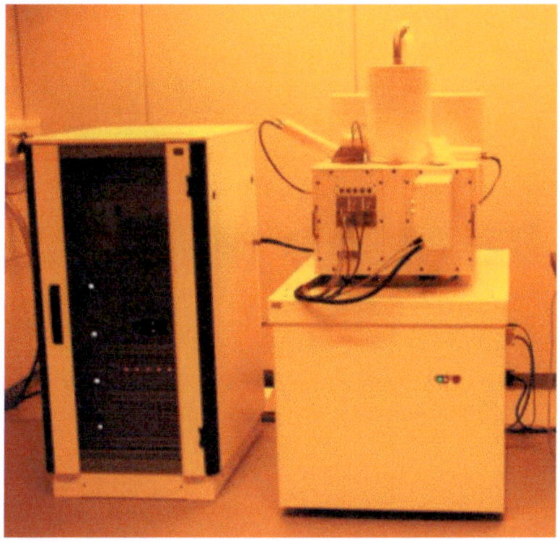

Figure 4.1: The Raith e_LiNE system as used for the nanostructure engineering

structures used throughout this work are to be investigated using optical means, so an optically transparent substrate is required. A thin glass cover slip is best for the use with oil immersion objective lenses for best imaging capabilities. Unfortunately, such a substrate is not conductive, which is a requirement for electron beam lithography processes as the electrons used to expose a resist need to be conducted away from the sample before it is too heavily charged. The charging problem is solved by the application of a very thin, transparent and conductive indium tin oxide (ITO) layer on the glass cover slip. Such a thin layer of ITO is sufficiently conductive to allow for electron beam lithography to be performed. Also, a lift-off process can be carried out at the end of the manufacturing to remove the resist without a removal of the desired gold structures. Thankfully, ITO does not heavily distort the optical measurements to be presented in later chapters. It is an insulator at optical frequencies and thus does not short-circuit two-arm antennas posed on top of it, and has a dielectric constant close enough to glass to allow for only very small changes of the optical response, and is reasonably transparent for thin layers.

A fabrication cycle for a sample with gold antennas is conducted as described in the following. The protocol is also summarized in Figure 4.2. First, a circular glass cover slip (24 mm diameter, PLANO) is baked out slightly below the glass melting temperature at 510 °Cfor about eight hours. This helps to further smoothen the glass surface due to surface healing effects and to remove non-glass residuals, thereby reducing sample auto-fluorescence when illuminated with highly intense laser light. Subsequently, ITO is sputtered

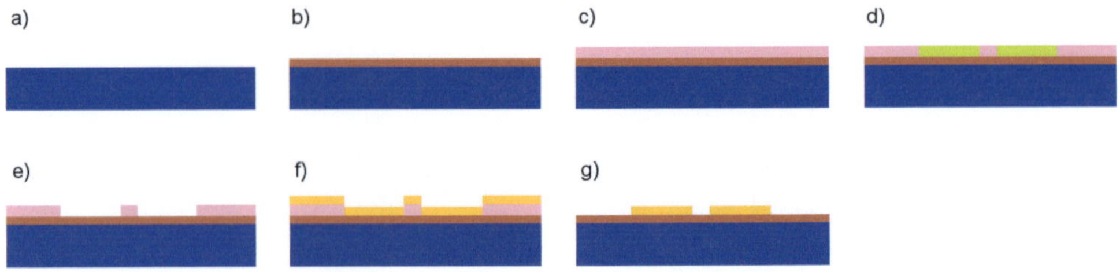

Figure 4.2: Gold antenna fabrication protocol: a) glass sample cleaning, b) ITO sputtering, c) PMMA spincoating, d) exposure, e) development, f) gold evaporation, g) lift-off. At the completion of the fabrication cycle, gold nanostructures have been prepared on a glass/ITO substrate.

onto the glass with a Univex 350 system (Leybold). A 30 nm layer of ITO is applied onto the sample after 150 s of sputtering at 100 W, with a vacuum pressure before starting the process of $8 \cdot 10^{-6}$ mbar and a working pressure of $1.5 \cdot 10^{-2}$ mbar.

After a standard cleaning procedure using acetone, isopropyl alcohol and an ultrasonic bath, a diluted PMMA (polymethyl methacrylate) resist (Allresist 950k, 1.1%) is spincoated onto the glass cover slip to be used for electron beam exposure. Spincoating is performed for 60 s at 8500 rpm, followed by a 300 s postbake at 175 °C. The resulting resist thickness was measured to be 44 nm.

Extensive dosage studies were performed to find a suitable set of electron beam lithography (EBL) parameters for the desired nanostructures. Details on these studies can be found in reference [188]. The lithography parameters that were found to be optimal for a reproducible fabrication of 20 nm wide optical antennas of variable length and gap size are listed in Table 4.1. Single exposure lines were used to write antenna structures, as the desired 20 nm width proved to be the thinnest reproducible fabrication width. A similar width could be achieved with area parameters as well, as will be shown in Chapter 4.2 on aluminum antenna fabrication.

Antennas could be arranged in any desired proximity and/or multitude in each write field, and several write fields could be placed on the same sample. Thus thousands of antennas could easily be fabricated on a single substrate. After lithography and development (using a standard developer, as well as isopropyl alcohol as stoppers), gold evaporation is performed in an ultra high vacuum turbopump system (Balzers 350). 30 nm of gold are evaporated at a pressure of 10^{-6} mbar and an evaporation rate of about 0.125 nm/s. To conclude the

parameter	value
line step size (LSS)	2 nm
line dwell time (LDT)	1.02 μs
beam speed (BS)	1.96 mm/s
aperture size	10 μm
write field (WF)	100 μm x 100 μm
high voltage	20 kV
current	29.4 pA
dosage	1050 pC/cm

Table 4.1: Exposure parameters for PMMA 950k 1.1% to generate a suitably exposed resist for gold nanostructures.

fabrication process, a lift-off is performed using acetone. The antenna sample generated can then be used with the optical characterization setup that will be described and used in the following chapters alongside the specific experiments.

4.2 Aluminum Antenna Fabrication on a Glass Substrate

As gold is not the only material of interest for optical antennas (and plasmonic structures and materials in general), a choice was made to also observe the optical characteristics of aluminum antennas. As seen in Chapter 2.2.1, aluminum has very different dielectric properties, which should yield different resonance frequencies compared to gold antennas. It was also decided to pursue a fabrication scheme directly on glass, as the ITO layer is in principle not required for the operation of the antenna and the sputtering process adds some additional surface roughness. A smooth surface is desired as it allows e.g. to perform nanomanipulation of particles on the surface, which may be used to selectively position single quantum dots near optical antennas at a later stage. So in this section, a method to avoid glass charging effects due to the missing ITO layer during the exposure procedure will be introduced as well. More details on the fabrication process than shown in the following paragraphs can be found in the *Studienarbeit* and *Diplomarbeit* by Patrick Schwab [165, 166] and the *Bachelorarbeit* by Ekkehart Schmidt [160].

The general fabrication process is very similar to the one outlined above, except that the sample that is to be used for the lithography process needs to be

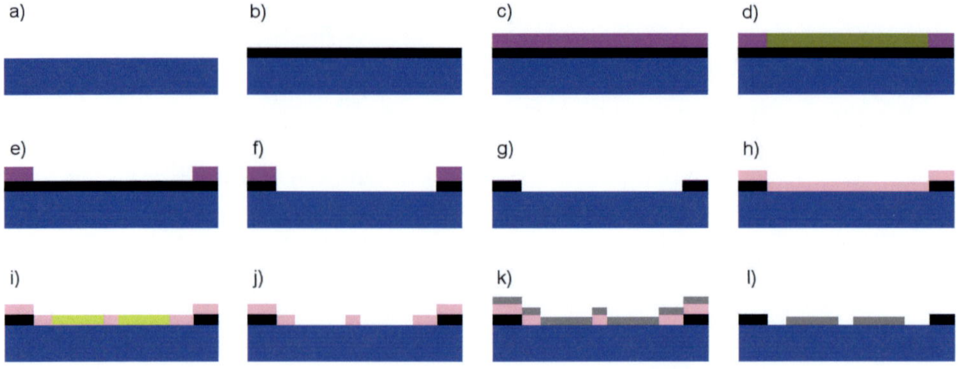

Figure 4.3: Aluminum antenna fabrication protocol: a) glass sample cleaning, b) niobium sputtering, c) photoresist spincoating, d) exposure using optical mask (in fact, an image reversal resist and thus also a reversed mask was used, this is omitted for simplification), e) development, f) niobium etching, g) stripping, h) PMMA spincoating, i) exposure, j) development, k) aluminum evaporation, l) lift-off . At the completion of the fabrication cycle, aluminum nanostructures have been prepared directly on a glass substrate.

prepared differently. The complete fabrication cycle from the first photoresist to the final lift-off process is detailed in Figure 4.3, a description is provided in the following.

To enable direct patterning for antennas on glass, an optical lithography mask was fabricated, which allowed structuring of the glass with a niobium layer and open glass windows. The layout of the structuring mask is shown in Figure 4.4. Besides the open glass windows, in which the antennas were to be written, several marker structures were also available to help to find specific sections of the sample both during the exposure procedure as well as during characterization. Since an image reversal resist was used, the structures to be removed from the resist and then from the niobium were non-transparent in the mask.

Niobium was used as a conductive material as it was readily available at the IMS. After the baking-out of the glass substrate and standard cleaning procedures using acetone and isopropyl alcohol, a thin layer niobium was applied to the substrate first. In principle, any other conductive material could be used, preferably with a low optical response to avoid optical crosstalk during characterization experiments. With a base pressure of $5 \cdot 10^{-3}$ mbar and a sputtering rate of 0.28 $\frac{\text{nm}}{\text{s}}$ for 90 s, a niobium layer of 25 nm thickness was applied to the sample.

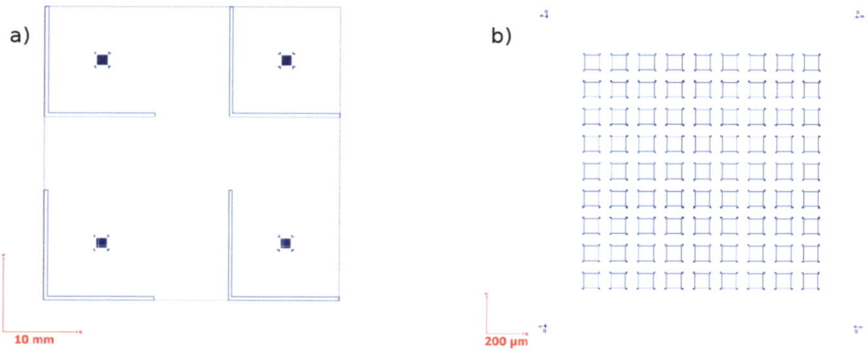

Figure 4.4: a) gives the entire optical lithography mask layout, with fields for four samples, b) shows the view of the central part of the layout for one sample, with 81 write fields of size 80 μm x 80 μm available for antennas.

After another cleaning procedure using n-hexane and acetone in an ultrasonic bath for 10 s each, a resist for optical structuring of the sample was applied. The image reversal photoresist AZ 5214E was used to obtain optical structuring. It was spin-coated on the sample at 7000 rpm (for 60 s), then, a prebake of 5 min at 85 °C was performed. Exposure using the mask shown in Figure 4.4 followed for 3.5 s in a Karl Süss Maskaligner MJB-3 system. Then, an image reversal bake step was performed for 5 min at 120 °C. Finally, flood exposure without any mask for 60 s was applied. Development was performed using the developer AR 300-47 MIF for 90 s (distilled water was used as a stopper), which led to some overexposure, but was long enough to ensure good removal quality for the desired window structures on the sample. Finally, the niobium was etched away through these windows using reactive ion etching. Some overetching was used to make sure the niobium was completely removed in the windows desired.

After the mask generation, PMMA was spincoated onto the sample as described in the previous section. The layer used needed to be slightly thicker, as the 44 nm layer that had been found to be optimal for the gold antenna fabrication proved to be too thin to avoid problems with the lift-off process for aluminum. Thus in this case a 1.5% concentration was applied with the same spin-coating parameters, leading to a resist thickness of about 60 nm. A 300 s postbake at 175 °C was used as before.

Lithography was then specifically performed within the glass windows on the substrate. The electron beam alignment was conducted only on the conductive material. A slightly different set of electron beam lithography parameters was used, adapted for the new resist thickness and the new substrate. Also, it

parameter	value
area step size (ASS)	2 nm
area dwell time (ADT)	0.122 μs
beam speed (BS)	16.351 mm/s
aperture size	10 μm
write field (WF)	100 μm x 100 μm
high voltage	10 kV
current	29.43 pA
dosage	126 μC/cm^2

Table 4.2: Exposure parameters for PMMA 950k 1.5% to generate a suitably exposed resist for aluminum nanostructures.

proved that the area mode instead of the line mode could now be applied. The parameters can be found in Table 4.2.

After exposure, again development, evaporation, and lift-off using acetone were performed. This time, the evaporation was carried out in a home-built thermal evaporation system at the IMS (mainly composed of parts fabricated by companies Leybold and Pfeifer). The system base pressure before evaporation started was approximately $1 \cdot 10^{-7}$ mbar, filament high voltage and filament current were 1.3 kV and 6.9 A, respectively. To achieve an aluminum layer (and thus antenna) thickness of again 30 nm, 24 s of evaporation at these settings were found to be required. After evaporation, the lift-off process was performed the same way as for the gold nanoantennas using acetone.

Again, many antennas could be fabricated on the same substrate, enabling extensive studies of optical antennas with the optical characterization setup that will be described alongside the specific characterization measurements in the subsequent chapters. Before that, a description of the topographical properties of the antennas that were fabricated will be given in the following section. The characterization of these properties is important, because only structures that are well known to have a specific shape can be correlated with confidence to their respective optical characterization results.

4.3 SEM and AFM Characterization

Scanning electron beam microscopy (SEM) is an easy and efficient tool to characterize the surface properties of optical antennas. In its standard form, however, it only gives 2-dimensional data, as the material contrast of the structure is used to discriminate it from its background. Nonetheless, it provides valuable insight, as one needs to be sure of a successful fabrication attempt before optical characterization can be performed.

The antenna fields used in this work were all designed such that each antenna could act independently from its neighbours. To this end, a minimal distance of 5 μm was typically kept between two structures. From each structure to the next, some antenna parameter was varied to allow for a quick optical characterization in a later step. The antenna spacing that was used for gold nanoantennas is demonstrated in Figure 4.5 with a scanning electron microscope (SEM) image.

Figure 4.5: Antenna spacing – SEM image of three gold antennas of different size as part of a 100 μm x 100 μm write field. Scale bar length is 1 μm.

The typical shape of a two-arm nanoantenna as observed using an SEM (top view) is shown in Figure 4.6. A few features are to be noted. First, the antenna shape is in good agreement with the design parameters. Evidently, however, the antenna structure is slightly curved on its edges, making the arms less rectangular than originally intended. This is due to the resolution limitations of eletron beam lithography at this very small scale. The finite gold and aluminum evaporation grain size also introduces some small errors. Such deviations, however, do not influence the observed scattering or plasmonic response significantly, such that only the nominal antenna parameters for width and length of the antennas will be used in the remainder of this work. Visible in the image is also a surface roughness of the ITO substrate, which also has some influences on the shape of the antenna itself.

Such antennas were also analyzed using atomic force microscopy (AFM) to yield height information not accessible using SEM alone. AFM images were taken using intermittent contact mode. A typical image that can be generated is given in Figure 4.7. The AFM image confirms the antenna height of about 30 nm and also describes the other physical dimensions nicely. The gap is reproduced faithfully in this image. In many measurements its size is not perfectly resolved using this technique (especially for very small structures), as AFM images are always a convolution of AFM tip diameter size (for the

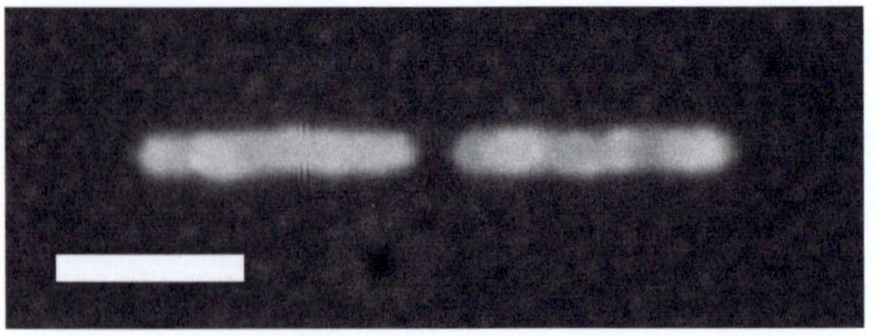

Figure 4.6: Typical SEM top view shape of a single two-arm gold nanoantenna fabricated on an ITO substrate. This antenna has a nominal arm length of 140 nm, antenna width of 20 nm, and gap size of 20 nm. Scale bar length is 100 nm.

tips used in this work their radius is about 10 nm) and the actual structure size. This also explains why the antenna looks slightly broader in the AFM image compared to the SEM one seen above. It should be noted as well that in the AFM image the ITO substrate roughness is also visible.

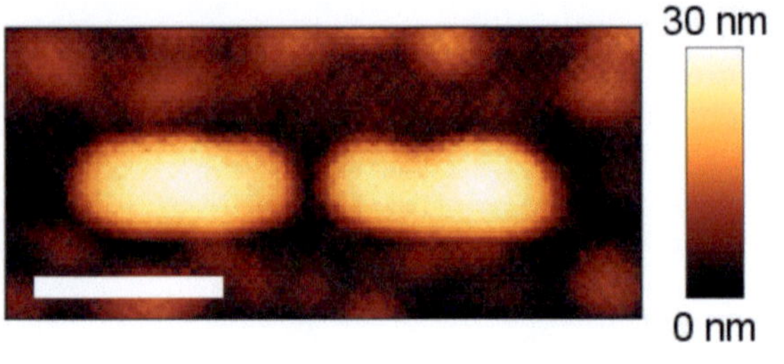

Figure 4.7: Typical AFM image of a gold two-arm optical antenna of nominal arm length 105 nm, nominal width of 20 nm, nominal height of 30 nm and nominal gap width of 20 nm as taken in intermittent contact mode. The false colour scale describes antenna height. The gap size is especially faithfailly reproduced in this image. Scale bar length is 100 nm.

For aluminum antennas, SEM and AFM images for structures of similar size look very similar to their gold counterparts. Notable is a slightly reduced contrast of aluminum on ITO compared to the images presented above for gold, which is simply due to the different material responses under the electron beam. SEM images of aluminum antennas on a glass substrate are very challenging to take, as the glass charging quickly destroys any imaging contrast. Only low resolution quick scans are possible. An improved imaging quality is reached on a niobium or silicon substrates. However, as already mentioned above, such substrates can not be used for optical characterization. For aluminum antennas on glass, the success of the fabrication protocol is thus generally established from an extrapolation of such antenna structures on

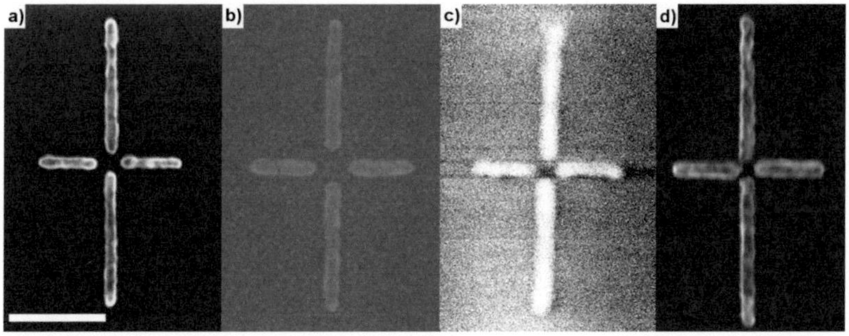

Figure 4.8: Four-arm aluminum antennas on different substrate types: a) silicon, b) ITO, c) glass, d) niobium. Scale bar length is 100 nm. The thinnest antenna arms and best imaging quality can be reached for a silicon substrate, which is, however, not useful for optical characterization purposes. The gap size difference in the images does not stem from the substrate type, but is due to different fabrication parameters used. (Images courtesy of Ekkehart Schmidt)

ITO-covered glass substrates. As seen in Figure 4.8 for four-arm aluminum antennas on different substrate types, this is, however, a very legitimate claim.

In the next chapter, the characterization setup developed and built to optically characterize the fabricated nanostructures will be presented. The following chapters will then describe the specific experiments performed in greater detail.

Chapter 5

Characterization Setup and Techniques: From Dark-Field Microscopy to Two-Photon Induced Antibunching

For the various experiments with optical antennas presented in the following chapters, a very flexible and multi-purpose characterization setup is required. Thus during the course of this thesis an experimental configuration was developed that allows to perform either simultaneously or with very short transformation times various optical excitation schemes, atomic-force microscopy measurements (as already presented in the previous chapter), detection of very low intensities down to single photons using avalanche photodiodes, and the recording of spectral information. In the following, the setup as well as the most important measurement techniques are described in an overview and the performance of the system is demonstrated using two examples: the measurement of the two-photon excitation point-spread function and antibunching measurements of single quantum dot luminescence under two-photon excitation.

5.1 Complete Characterization Setup

The characterization setup in its final form as developed throughout this thesis is shown in Figure 5.1. It allows for several experimental techniques to be applied, among them dark-field microscopy in transmission and reflection, confocal microscopy, and photon correlation measurements including photon antibunching investigations. Centerpiece of the setup is a Zeiss Axiovert Observer microscope including a JPK Instruments atomic force microscope. A three axis piezo stage holds a sample under investigation and can move it in all three dimensions. For optical excitation purposes, several laser and white-light sources can be used applying switching mirrors, while for detection either avalanche photodiode (APD) detectors, a real colour imaging CCD or a spectrometer with an attached EMCCD or nitrogen cooled CCD detector can be applied. Several wavelength filters and polarization filters can be used to select specific excitation conditions and/or separate the excitation light from the response of the structures. The specific excitation sources, optical elements, and detectors will be named in the following sections, where the measurement techniques outlined above are described in more detail.

5.2 Dark-Field Microscopy

Dark-field microscopy is the technique of choice to investigate linear scattering properties of nanostructures, since it allows to single out the scattered light (for example by an optical antenna) from the light used for excitation. Obviously, for best results a broad-band (or widely tunable) light source needs to be used, as only wavelengths of light that are incident on the antenna can be scattered. Dark-field microscopy can be used as an imaging or spectroscopic technique, depending on the results desired. The general idea is rather simple: If one illuminates a sample which includes scattering particles, and then finds a way to filter the light that was used for the illumination, bright scattering spots can be observed using a microscope. However, elastic scattering as observed in this work is a process that conserves wavelength, so frequency filtering (as commonly done for fluorescence microscopy) cannot be used. One must thus resort to spatial filtering, which can be performed in several ways. Two configurations are most commonly considered as typical dark-field microscopes, which are presented in Figure 5.2. Common to both the setup in reflection and the one in transmission is that only a ring of light is used to illuminate the sample. For the transmission setup, an objective lens of lower numerical aperture than the innermost components of that ring is then used to

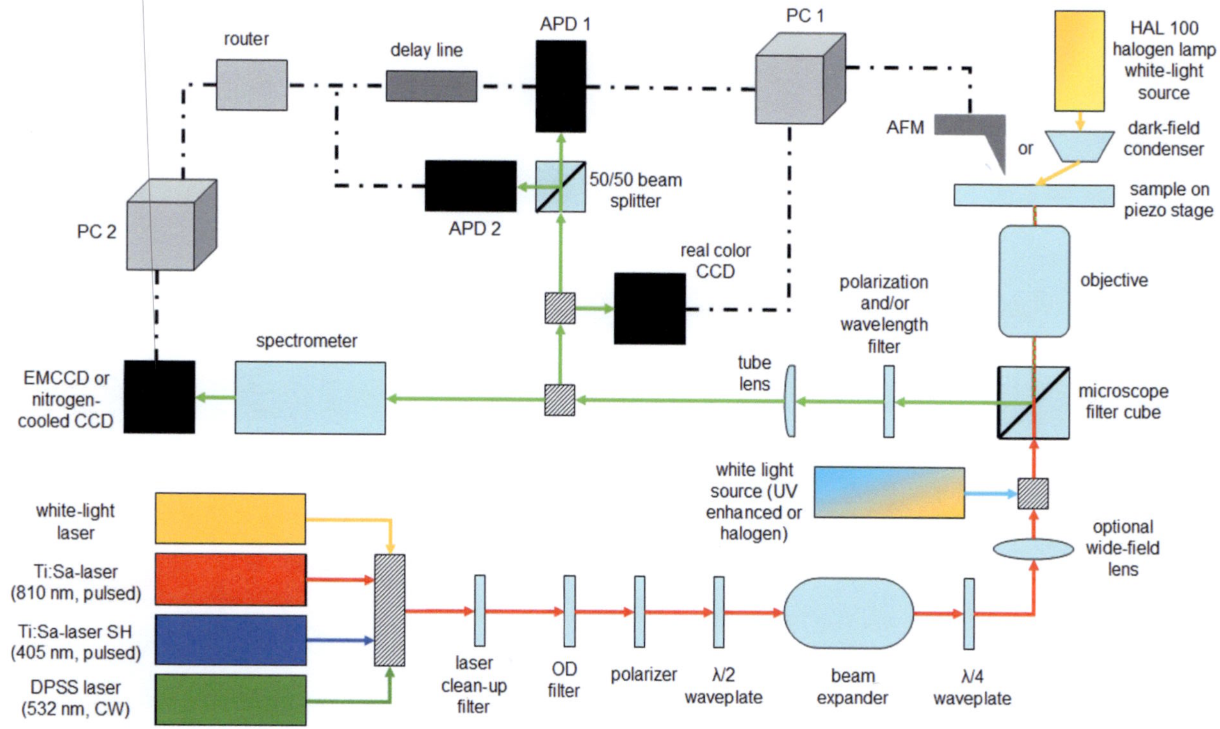

Figure 5.1: The characterization setup built for the experiments carried out in the following chapters. Light sources are given in colour, detectors in black. Optical paths are depicted using straight lines, while electrical data transfer is visualized using dot-dashed lines. Boxes with diagonal stripes are points where manual selections between excitation sources or detectors need to be performed.

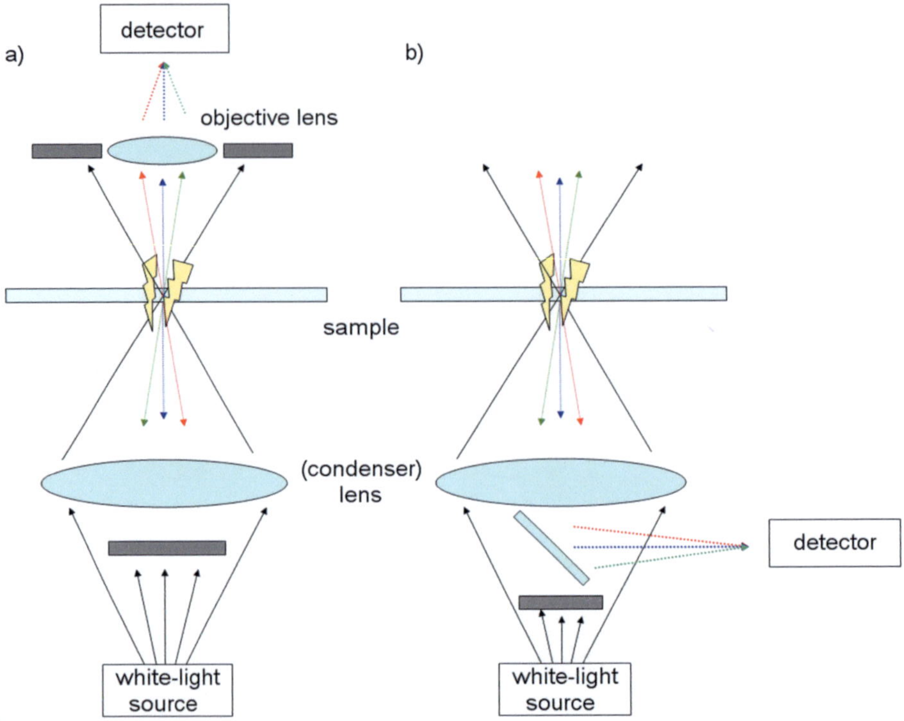

Figure 5.2: Dark-field microscopy in a) transmission and b) reflection configuration. The white-light continuum (e.g. from an ordinary light bulb) used for excitation is shown with dark lines, while the scattered light generated on the sample at the position with the lightning bolts is depicted in red, green, and blue.

collect the scattered light exclusively. This numerical aperture mismatch thus serves as the filter for the excitation light. In the reflection setup, the same numerical aperture objective lens is used for excitation and collection. For the collection, it is e.g. possible to only forward the inner light cone by the use of a tilted mirror. In many setups, it is the excitation light which is incident under a ninety degree angle, then the mirror in the detection path can be replaced by simple free space. No matter which setup is chosen, points from which no light is scattered appear dark in the image obtained – this is why dark-field microscopy bears its name. Bright structures appear in a field of darkness. As it is a far-field investigation method with a broad-band light source, its resolution is diffraction-limited.

Surprisingly, even though the method is nowadays in common use for the characterization of nanostructures down to a size of a few tens of nanometers (for smaller structures absorption cross sections tend to overwhelm the scattering properties), the invention of the method is hard to track. While the invention of ultramicroscopy, which is also based on light scattering from particles, but with a somewhat different excitation scheme, can be easily attributed to Richard Adolf Zsigmondy, reliable sources pinpointing the invention of the dark-field microscope to a specific researcher could not be found. (Claims that the invention was made by August Köhler around the beginning of the 20th century may be correct, but lack proper citations).

To describe the power, but also the beauty of dark-field microscopy, Figure 5.3 is an example for an image obtained using that technique. Several gold colloids of approximate diameter of 100 nm were drop casted on a glass cover slip and then investigated using dark-field microscopy applied in reflection mode. Each single colloidal particle yields a scattering response according to the particle size (which can be described using Mie's scattering equations). Due to small size variations, the apparent colour changes from green to orange for different particles. When taking spectra of such structures, great care has to be taken to acknowledge the spectral properties of the lamp used for the excitation of the scattering response. Any non-uniformity in the spectral range directly transforms to a change in the measured spectrum. As the relationship is linear, however, this problem can be overcome by normalization of the excitation light source and subsequent division of the measured signal by the obtained curve. The image convincingly demonstrates that dark-field microscopy can be easily applied to investigate gold nanostructures.

In the measurement setup, dark-field microscopy in transmission is performed as follows: The built-in lamp of the Zeiss Axiovert Observer microscope (Zeiss HAL 100, colour temperature 3200 K, with its IR-filter removed to also allow for longer wavelength responses) homogeneously illuminates a dark-field condenser (Leitz Wetzlar, NA = 1.2), which is placed in immersion oil on top of the sample. The scattered light from the structures on the back of the sample is then collected using a second objective lens (Zeiss 40x, NA = 0.95). The condenser is placed on the sample such that an area of about 1 mm^2 is illuminated. Accordingly, a write-field of size 100 μm x 100 μm containing an array of antennas can assumed to be illuminated homogeneously. The collected scattering light is then either imaged using the real colour CCD camera or the electron-multiplying CCD (EMCCD) camera (Andor iXon). The latter can also be used in conjunction with a spectrometer (Acton 2500i). The spectrometer is equipped with several gratings; for all

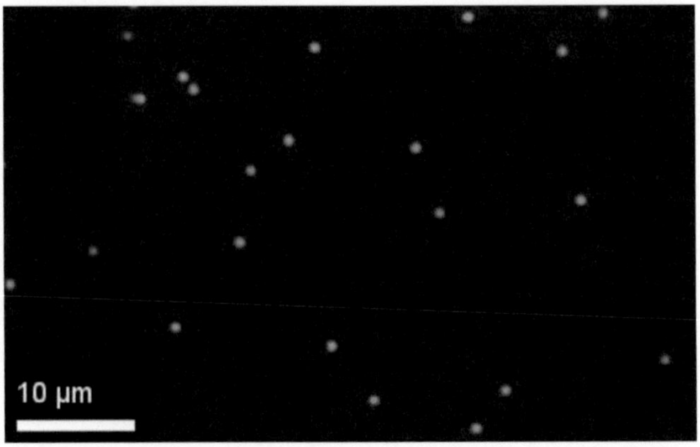

Figure 5.3: Real colour CCD camera image of 100 nm diameter gold nanoparticles (nominal size) on a glass substrate, imaged using dark-field microscopy in reflection with a dark-field objective lens. Slight size variations of the single particles yield colour variations in the scattering response.

experiments described later a 500 nm blazed, 300 lines/mm grating is used, with which a response over a spectral range of about 300 nm can be observed. For greater range, several images can also be stitched. Dark-field microscopy in reflection is also possible with the setup. To this end, a white-light source in front of the microscope filter cube is used. Then, a special dark-field objective lens both generates the ring-shaped excitation light and properly filters it out for detection. This is important for possible combinations with AFM measurements, as the condenser required has to be placed in the same place as the atomic force microscope head, as also visualized in Figure 5.1.

5.3 Confocal Microscopy

Confocal microscopy was invented by Marvin Minsky, who applied for a patent in 1957 [124]. His aim was to overcome the cross-talk often seen in wide-field fluorescence imaging: If a sample is homogeneously illuminated with excitation light, then all fluorescent specimen on the sample surface will show luminescence, which makes the discrimination of features challenging. In confocal microscopy as depicted in Figure 5.4, a point-like light source (typically a laser, often sent through an additional very small aperture) is focused on a sample. The luminescence from the sample is collected, sent through a second aperture in an image plane and then detected. The aperture in the image plane is conjugate to the aperture in the excitation channel, which is why confocal microscopy bears its name. Due to the spatial filtering, only luminescence from the optical focus is collected, which under ideal circumstances yields an improved resolution by a factor of $\approx \sqrt{2}$ over wide-field fluorescence imaging [147].

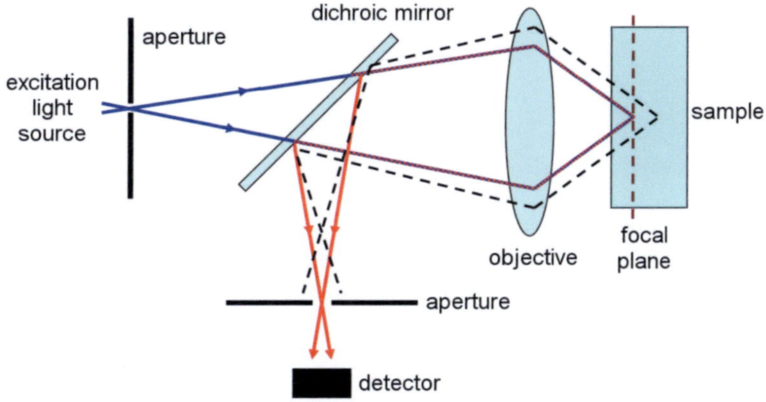

Figure 5.4: Confocal microscopy in reflection, using a dichroic mirror to split the luminescence signal from the excitation light. Light emanating from outside the focal area (dashed black lines) is significantly reduced in intensity by the aperture in front of the detector. (Image inspired by Minsky [124])

In the setup outlined in Figure 5.1, each of the laser sources depicted in the lower left corner can be used for excitation in a confocal microscopy setup. A pulsed (80 MHz) white-light supercontinuum laser source (Koheras SuperK Versa), a pulsed Titan-Sapphire laser (76 MHz) typically operated at 810 nm (Coherent Mira), its second harmonic signal at 405 nm, and a CW diode pumped solid state laser (HB-Lasers DPSS Green 532) are available. Their extent is assumed to be sufficiently point-like for confocal microscopy, so no additional aperture is introduced in the excitation channel. The beam intensity can be reduced using neutral density filters placed in two filter wheels (Thorlabs FW2AND). The linear laser polarization is enhanced using a VIS Glan-Thompson Polarizer (New Focus 5524). A $\frac{\lambda}{2}$- and a $\frac{\lambda}{4}$-waveplate (New Focus 5540) are available to set a desired polarization state. A beam-expander comprising Thorlabs lenses finally increases the beam width for a back lens overfill of the objective lens by a factor of 2.5. The oil immersion objective lens (Zeiss Plan-APOCHROMAT, NA = 1.46) is used with Immersol 518F

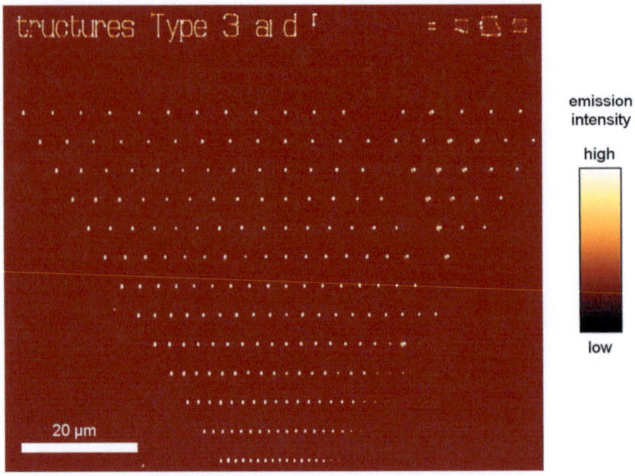

Figure 5.5: A 2D two-photon luminescence raster scan of an array of optical antennas as an example for the raster scanning procedure available using the characterization setup.

(n=1.518) immersion oil. Additional filters can be added at various positions, they will be introduced alongside the respective experiments in later chapters.

The intensity from the spot under investigation can be imaged on the real colour CCD camera or investigated spectroscopically using an Acton 2500i spectrometer with the attached Andor EMCCD or Jobin-Yvon CCD. For confocal resolution, however, it is sent towards two avalanche photodiodes (APD, EG&G). The APD-detectors are connected to a photon counting card (SPC830, Becker & Hickl), which enables digital read-out of the single photon stream. As they are placed in an image plane, the small extent of the APD detector area acts as the detection pinhole of the confocal setup. For the confocal experiments, only one APD is required, the need for the second detector will be explained in Chapter 5.4.

The APDs also allow for 2D raster scans of parts of the sample. To this end, the sample can be raster-scanned in the plane perpendicular to the optical axis using a tip-assisted optics (TAO) piezo stage (JPK Instruments). For an image, the sample is divided into a number of pixels, which are examined one by one. The sum of the photon counts generated at the detectors for each pixel then yields a 2D image. An example is shown in Figure 5.5, where a write-field containing several antennas of different sizes has been excited using two-photon absorption. More details on the procedure can be found in Chapter 7, where the two-photon luminescence of antenna structures is thoroughly investigated. To obtain a raster scan of this quality and resolution, about one hour is required.

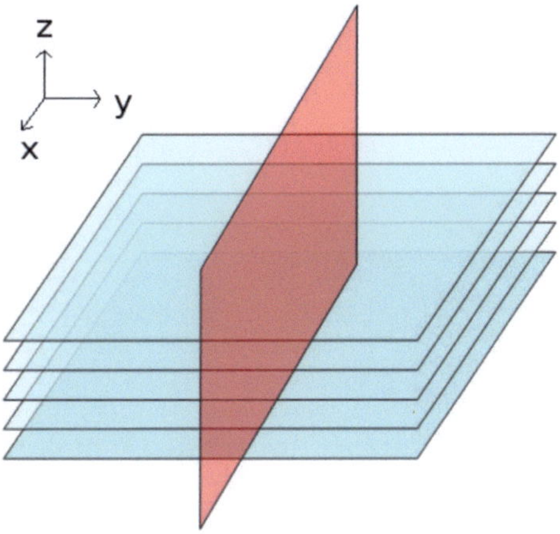

Figure 5.6: Several scans are performed using the x-y-axis stage piezo crystals (light blue). The images are then combined to yield a 3D matrix, from which a calculated image in the third dimension (here named z) can be obtained (light red).

Several such 2D images taken with small differences in height also controlled using piezo crystals can be stacked to obtain a 3D scan. From the equally spaced 2D images, z-information can be interpolated, as depicted in Figure 5.6: Several light blue images are measured in different planes (changing in this case the z-axis piezo position between scans), a cut through the planes can then be interpolated as a new image, yielding a plot along the third axis.

For the non-linear emission property investigations of the optical antennas in the following chapters, two-photon absorption confocal microscopy using the pulsed Ti:Sa laser operated at 810 nm is the technique of choice. It is thus important to know the optical resolution that can be obtained using the raster scans. This resolution limit as obtained by raster-scanning a known point-like luminescent object is also known as point spread function (PSF) and will be determined using a quantum dot in the following section.

5.3.1 Point Spread Function

A single quantum dot of the size of a few nanometers can be used as a stable chromophore to measure the point spread function of a microscopy system, where it is considered a point-like probe for the apparent interaction focus. To show that the optical characterization system operates successfully as a confo-

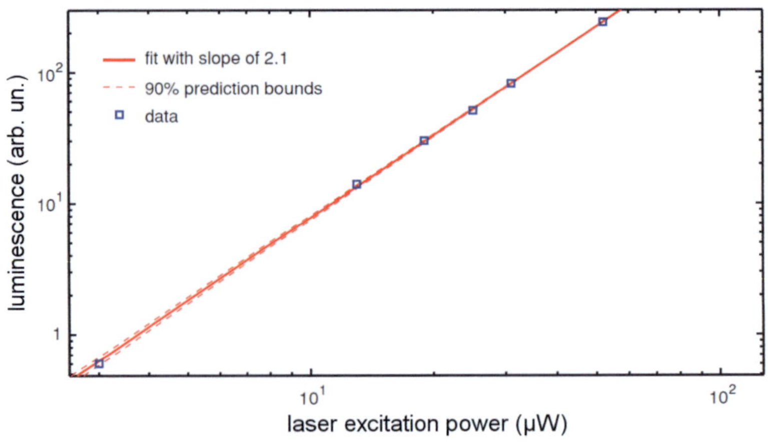

Figure 5.7: Double logarithmic plot of the laser excitation power *versus* the PL luminescence rate showing a slope of 2.1±0.1.

cal microscope, the two-photon[1] excitation-cross section will be mapped out using a single quantum dot as a chromophore. The contents of this section have already been published as a part of a publication on antibunching of quantum dot luminescence under two-photon excitation in *Physical Review B* [192].

To obtain the point-spread function, CdSe nanocrystals in solution emitting photoluminescence with a peak wavelength around 605 nm and an ensemble relative quantum yield between 40 and 50% were used. They were diluted to 10^{-10} molar and then drop cast onto a clean standard glass cover slip (thickness 170 μm). The glass sample was covered with an additional PMMA layer (approximately 100 nm thick) on top to immobilize the quantum dots.

In Figure 5.7, the quantum dot photon luminescence intensity is plotted *versus* the laser excitation power. With an average laser power I from 3 μW to about 50 μW at 810 nm (at 76 MHz laser repetition rate and an approximate pulse width of 500 fs at the laser focus) applied to a single quantum dot, a PL rate R that scales nicely as a square-dependence was measured. The rate equation $R = R_\infty \frac{(\frac{I}{I_{sat}})^2}{1+(\frac{I}{I_{sat}})^2}$ is thus applied in a range where the average laser power used is clearly well below the saturation regime $I_{sat} \stackrel{def}{=} \frac{1}{2}R_\infty$. For easy measurement of the curve, it is also beneficial that those semiconductor nanocrystals have among the available 2PE chromophore systems the highest 2PE cross section measured so far [106]. This relaxes the experimental fine adjustment

[1]On a side note, two-photon absorption had been theoretically described by Maria Göppert-Mayer as early as in 1931 [73], but the field of two-photon excitation microscopy was only opened by W. Denk et al. in 1990 [46].

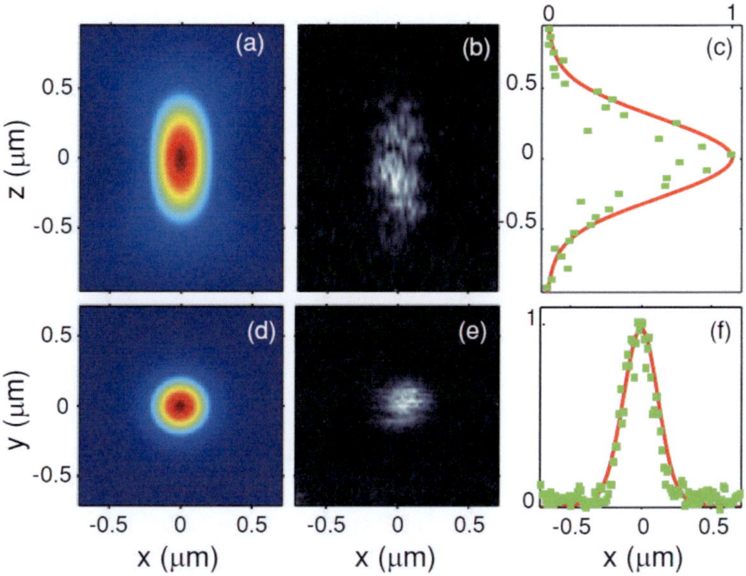

Figure 5.8: a) and d) show the model of the confocal point spread function (PSF); b) and e) show the confocal point spread function (PSF) measured with a single quantum dot as the probe; c) and f) show the comparison (normalized) of the model PSF (solid line; line cut through center of a) and d)) with the experimental PSF (scattered points; cut through center of b) and e)).

for improving 2PE induced PL generation efficiency by pulse compression devices near the focal region.

It has been shown experimentally that 2PE cross sections on the order of 66,000 GM[2] for CdSe quantum dots at room temperature easily start to saturate the confocal detection volume, which yields a confocal volume that is largely extended [106]. It was thus taken great care to only apply a laser average power well below saturation I_{sat} to measure the confocal point spread function of the optical system with a single quantum dot as a local delta type probe. To achieve this, the three axes piezo stage available was used to generate a stack of raster images of the optical PL response of a single quantum dot to the excitation laser focal volume. Here, images of size 1.5 μm x 1.5 μm were taken in 50 nm steps and combined to yield images for the x,y as well as the x,z-plane (where z is the direction of light propagation). The result is shown in Figure 5.8 b) and e). The quantum dots are extremely stable over time, no degradation is observed throughout the data acquisition time (< 1.5 h). There

[2]Göppert-Mayer units, 1 GM = 10^{-50}scm^{-4}photon^{-1}molecule^{-1}

is, however, some blinking of the quantum dot [135], which gives the images a slightly fractured appearance.

The measured point spread function under 2PE conditions is now compared with a model using a kernel based on an algorithm for a fast focus field calculation [109]. The modeling takes into account evanescent field contributions for specifically polarized input fields, as well as all the pertaining optical parameters – high NA, effective refractive index of the quantum dot embedding medium $n_{eff} = 1.5$, the polarization state of the incoming laser, the excitation wavelength, the emission wavelength, the Gaussian beam shape, and microscope objective overfill. The data was kindly provided by Hans-Jürgen Eisler. The model predicts a FWHM in the x,y-plane near the focus of ≈ 285 nm and in the x,z-plane of ≈ 657 nm, which is depicted in Figure 5.8 a) and d). A comparison of the measured data with the modeling via central cuts through the respective images is shown in Figure 5.8 c) and f). Model and measurement agree nicely here, so it can be stated that the major contribution to the confocal point spread function stems from the intensity square-dependency of the excitation point spread function. Thus in fact the two-photon excitation PSF has been mapped out and the characterization setup is known to operate at the limit of the resolution capabilities of confocal microscopy.

The APD detector in conjunction with the counting card as used for the raster scan imaging not only counts photons, but also pinpoints those photons to the moment in time in which they are detected in comparison to a reference time generated by a pulsed laser. This can be used to perform time-correlated single photon counting, as described in the next section.

5.4 Time-Correlated Single Photon Counting

Even though lifetime measurements and auto-correlation measurements will not be applied to optical antennas in the following chapter, it is interesting to shortly describe the possibility of the setup to perform time-correlated single photon counting (TCSPC) measurements. This is especially useful if in the future antennas can be combined with single quantum dots and lifetime information and emission auto-correlation functions of the hybrid structures are desired. The measurements presented in this and the following section have been published in *Physical Review B* [192] alongside the 3D point spread function measurements shown above.

The APD detectors already described earlier send an electrical pulse to the counting card in the measurement PC. When a pulsed laser is used to generate the signal, the laser pulses can also be detected using a high speed photodiode module (PHD-400, Becker & Hickl), by which the pulse position in time (and thus the laser frequency) is also fed to the counting card.

To allow for the measurement of lifetime information, the time between a laser pulse and a generated photon is measured. As the laser pulse is repeated at a very high frequency, and not each pulse generates a photon, a memory-saving way of measuring the time between photons is to measure the time between a detected photon and the next laser pulse event, which is then subtracted from the time between laser pulses to generate the decay time of the particular event observed. The measurement is digital, meaning that the time interval observed is distributed on a number N of time-channels of equal length. When many such measurements are taken, a statistical representation of the lifetime of a fluorescent event is generated.

If a raster scan is performed slowly enough such that a sufficient amount of photons [97] is generated for each pixel, the lifetime information can be obtained for each of them, a procedure named fluorescence lifetime imaging (FLIM). Figure 5.9 depicts such a measurement under two-photon excitation.

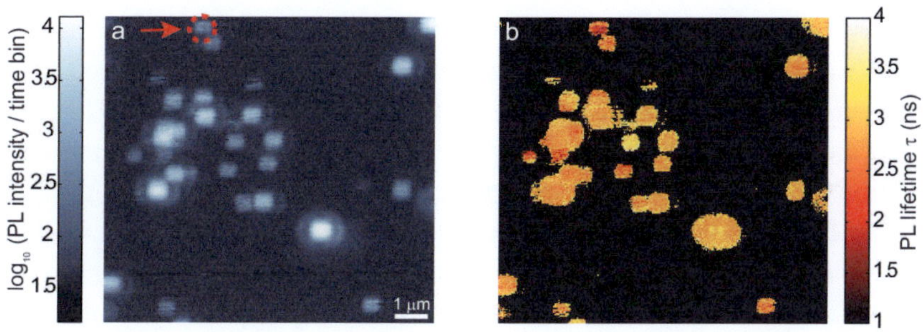

Figure 5.9: a) intensity information as obtained from two-photon excitation and a raster scan over an area of 10 μm x 10 μm, b) corresponding lifetime information for each pixel and thus the different quantum dots. The quantum dot singled out by the arrow and the dashed circle in a) will be investigated more closely in Chapter 5.4.1.

Several quantum dots (same preparation as for the point-spread-function measurement in the previous section) are present in a region of 10 μm x 10 μm. In Figure 5.9 a), the intensity is given, while 5.9 b) gives the corresponding lifetime information for each pixel. Local differences in the environment, dipole orientation and individual photoluminescence quantum yield from quantum dot to quantum dot yield a lifetime information contrast for various quantum dots with an excited-state apparent lifetime mean for all quantum dots in the scan area of $\langle\tau\rangle_{\text{mean}} = 2.8$ ns.

With the addition of a second detector and a 50/50 beam splitter before them, the auto-correlation of a specific signal can also be measured, which is commonly known as Hanbury Brown-Twiss configuration [76, 78, 77]. To this end, one of the detectors can be used as the second input of the photon counting card instead of the laser signal, and the time between two photon counting events is detected. Unfortunately, the photon counting card has a detection dead time of about 100 to 125 ns [12], which does not allow for correlation times below that value. Fortunately, this problem can easily be overcome by the introduction of a delay line (a long cable) after one of the detectors. The correlation at the orginal point of time 0 is then found at the additional time introduced by the delay line, which in the setup presented in Figure 5.1 is 250 ns. The correlation signal is again displayed as a digital data stream, where within a long time-frame the frequency at which a certain time difference between the two detectors to yield a photon is observed.

With this hardware correlation setup, the lifetime information as obtained from the comparison with the laser signal is no more available, which is why a software correlation scheme is desirable. An additional advantage is that with a software approach no changes to the hardware setup are required. To use software correlation, the two detectors are connected to a router, which gives each photon counting event the additional information whether it came from detector one or two. Then, the photon stream is connected to one of the entry points of the counting card, while the laser frequency is again fed to the first connector. Again with the delay line introduced for one of the detectors a stream of photons is recorded for a longer time frame for both detectors. The two photon streams can then be correlated using software. This has been performed to obtain the auto-correlation of a single quantum dot under two-photon excitation and will be presented in the following.

5.4.1 Two-Photon Excitation Induced Antibunching of Single Quantum Dots

A quantum dot as modeled by a two-level system of a ground and an excited state can only emit one photon at a given time (neglecting bi-excitonic emission, the next photon can only be sent out once the system has been excited again) – it can thus be viewed as a single photon source. The behaviour of only sending out one (or no) photon at a time is known as photon *antibunching*. Mathematically speaking, if antibunching occurs, then the auto-correlation measured for a single quantum dot is zero at $\tau_{\text{inter}} = 0$. τ_{inter} describes the time between two subsequent photon detection events, which corresponds to

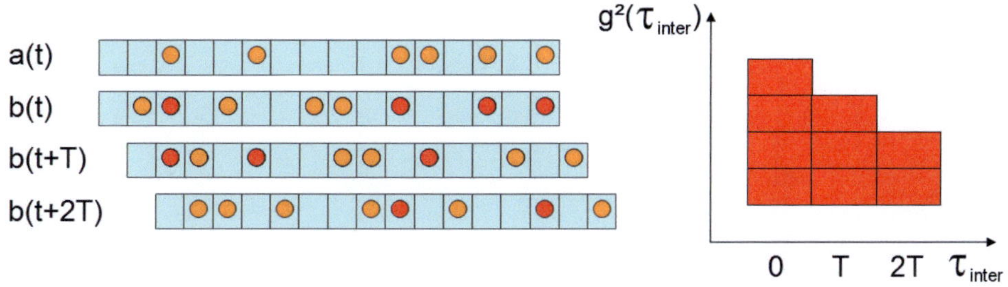

Figure 5.10: Algorithm to obtain the cross-correlation function from binary data. For each value of τ_{inter}, e.g. τ_{inter} = T, the coincidences between a detection event in photon streams a and b (photons marked by dots) at specific time intervals are counted (red dots) and summed up.

a specific shifting time for the correlation function. For a larger number N_{QD} of emitters, it can also be shown [146] that the auto-correlation $g^2(\tau_{\text{inter}})$[3] at $\tau_{\text{inter}} = 0$ becomes

$$g^2_{\text{norm}}(0) = 1 - \frac{1}{N_{QD}}. \tag{5.1}$$

As already mentioned above, the auto-correlation is measured by splitting the signal emitted by the source onto two detectors by means of a 50/50 beam splitter and determining the cross-correlation from the two detectors. The (not yet normalized) cross-correlation of digital data where for each time bin either one or no photon is recorded in two sets of data streams $a(t)$ and $b(t)$ from two detectors can be calculated using

$$g^2(\tau_{\text{inter}}) = \sum a(t)b(t+\tau_{\text{inter}}) \tag{5.2}$$

if the time bins are equidistant. This procedure is visualized in Figure 5.10.

To perform this operation, the counting events as recorded in the photon stream from each detector need to be transformed to a new time frame with constant binning intervals, as this is not necessarily the case for the original data. Then, shifting according to the above equation can be performed to obtain correlation data for a specific τ_{inter}. In the setup used the time lag from the delay line is fixed to 250 ns, so only correlations around that value need to be calculated. A bin resolution of 100 ps was used. This requires about 10 minutes to generate 120 ns (that is 1200 data points) of the correlation curve

[3] In the following, the auto-correlation is measured using two detectors, which means mathematically a cross-correlation is calculated. $g^2(\tau_{\text{inter}})$ will be used as the mathematical notation for both cases.

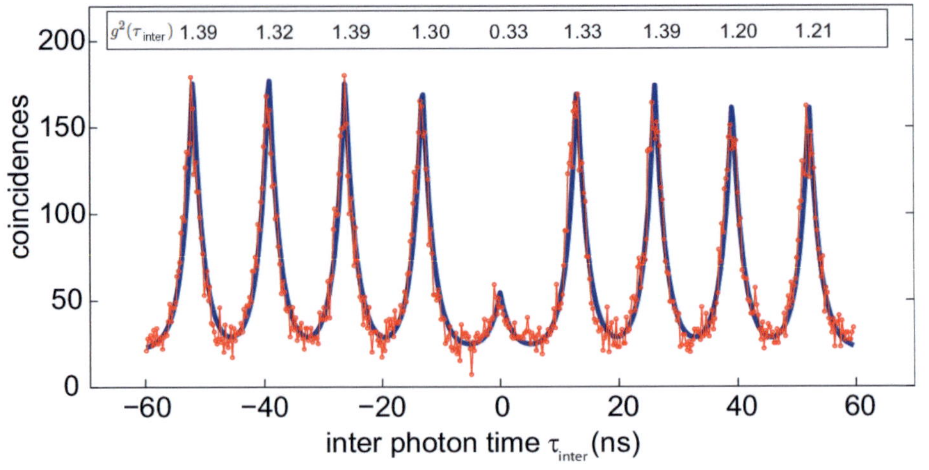

Figure 5.11: The measured auto-correlation curve under two-photon excitation of a single quantum dot around $\tau_{\text{inter}} = 0$. More details are found in the text.

around $\tau_{\text{inter}} = 0$, with each photon stream containing about 300 million photon counts (typical count rates of $5 \cdot 10^5 \frac{1}{\text{s}}$).

The correlation curve can be normalized to the total counts as expected from equal distribution, using [60]

$$g^2(\tau_{\text{inter}})_{\text{norm}} = g^2(\tau_{\text{inter}}) \cdot \frac{N_{\text{timeframes}}}{N_{\text{phot,det 1}} \cdot N_{\text{phot,det 2}}}, \tag{5.3}$$

which is required to obtain the correct values to check for the properties of a single photon source as defined in Equation (5.1).

Using this software correlation technique, antibunching behaviour of a colloidal CdSe quantum dot at room temperature was observed from the quantum dot surrounded by a dashed circle in Figure 5.9 a). The single quantum dot was put into the confocal laser excitation volume, and photon streams were measured using the two APDs for 10 minutes. Then, software correlation was applied as described above, with the shifting time 100 ps and a correlation interval observed around the known delay line offset of 250 ns of ±60 ns. The resulting curve is depicted in Figure 5.11. The reduced peak height at $\tau_{\text{inter}} = 0$ directly shows that antibunching behaviour is observed.

The coincidence maxima in Figure 5.11 derive from the excitation with a pulsed laser. The distance between peaks therefore yields the repetition frequency of 76 MHz. It is apparent that the correlation function does not drop all the way down to the zero level in between the coincidence maxima. This inter photon signal background is tentatively explained by coincidence contribution from the substrate due to non-perfect spectral filtering under the pulsed high peak power in the focal region, the intrinsic detector background noise level as well as the finite time resolution of the experiment of 100 ps. Furthermore, for this size of CdSe nanocrystals, photon pair coincidence contributions from multi-exciton radiative relaxations for $\tau_{\text{inter}} = 0$ cannot be entirely ruled out [62].

The normalized intensity auto-correlation function values $g^2(\tau_{\text{inter}})$ are obtained as described in Equation (5.3). The normalized peak values of $g^2(\tau_{\text{inter}})$ > 1 at $\tau_{\text{inter}} \neq 0$ points towards the well-known additional bunching effect[4] mediated by the meta-stable quantum dot blinking states (on-/off-states). Consequently, quantum dot blinking or intermittency shows no characteristic time scale over which a photoluminescence intensity average for nanocrystals can be given – they are non-ergodic [122].

As $g^2(\tau_{\text{inter}} = 0) = 0.33 < 0.5$ (which would be the value as obtained for two quantum dots from Equation (5.1)), it is clear that a single quantum dot is present in the 3D focal volume, even without correction for the background level. The small peak at $\tau_{\text{inter}} = 0$ may be due to PMMA background luminescence, as already observed by other groups [180].

The peaks in the correlation function were also fitted using a sum of single exponentials with a least-square algorithm, taking into account the photon coincidence background. From the fits, the apparent excited-state lifetime of the freely emitting quantum dot is found to be $\tau = 1.7$ ns. This is the same value as already obtained for that particular quantum dot in Figure 5.9 a).

5.5 Further Capabilities and Conclusions

The characterization setup has many additional capabilities. Beside the use of the atomic force microscope, also in conjunction with optical excitation (as presented later as a newly developed measurement scheme in Chapter 8), the possibility to perform the nanomanipulation of colloids to obtain specific patterns (more details can be found in the *Studienarbeit* by Fabian Herberger [81]) and confocal dark-field imaging (as developed in the *Bachelorarbeit* by William Truong [177]) are probably the most promising. All of these techniques have potential applications for optical antennas. For example, the controlled nanomanipulation of gold colloids can offer the possibility to engineer antennas made of two or more colloids with various gap sizes.

[4]Photon bunching, i.e. an enhanced statistic probability of photons to arrive at a detector simultaneously, is e.g. observed for thermal light sources.

In summary, this chapter has outlined the characterization setup that was developed throughout the thesis and highlighted the two experimental techniques most crucial for the following chapters, namely dark-field microscopy and confocal microscopy. In addition, the measurement of lifetime information and antibunching from two-photon excitation was presented, which is a useful tool for future combinations of optical antennas with such quantum dot emitters, as then again the single-photon emission capabilities of a potential hybrid device can be observed. The following chapters will now present the experimental findings on the optical characterization of nanoantennas, starting with their linear scattering properties in Chapter 6.

Chapter 6

Dark-Field Scattering Characterization

In this chapter, the scattering properties of optical antennas, that is their resonance behaviour, will be investigated. In the following, the dark-field microscopy setup in transmission as outlined in Chapter 5.2 will be used to characterize optical antennas with two or four arms. Parts of this chapter concerning the scattering of gold nanoantennas have already been published in the journal *Nanotechnology* [193].

Typical integration times for the scattering spectra presented in the following were about 150 s (with the electron multiplication feature of the EMCCD camera turned on). All spectra were corrected for the spectral shape of the excitation light. The changes introduced by this correction are fairly weak, as the spectrum of the halogen lamp used is comparably flat. A polarizer could be introduced into the detection channel, which enables polarization selective measurements.

6.1 Two-Arm and Single-Arm Gold Antennas

As stated in the previous sections, optical gold nanoantennas need to be sufficiently small and even shorter in length than the classically well-known $\frac{\lambda}{2}$ limit. The exact scattering wavelength response, however, may also be influenced by parameters such as gap size (and therefore particle coupling). To find out more about the gold nanoantenna scattering, it was therefore decided to systematically vary two antenna parameters for the dark-field scattering experiments: antenna length and antenna gap size (for two-arm dipole antennas). For comparison, single-arm antennas without gap were also investigated.

6.1.1 Antenna Arm Length Variation

For the length variation, to allow for a quick comparison and the measurement of several antennas at the same time, the antenna length variations were performed in two lines with nine two-arm antennas each. Starting with 25 nm, the arm length L_{arm} on each side was increased by 5 nm for each antenna, yielding a first line with L_{arm} from 25 nm through 65 nm, and a second line with L_{arm} from 70 nm to 110 nm. The arm width was measured to be 20 nm using SEM images and the height confirmed to be about 30 nm, as shown in the previous section. The gap size for two-arm antennas was kept constant at 20 nm. The arm lengths were chosen according to previous publications [128, 133] which showed that a response spectrally located in the visible can be expected in that antenna size regime, even though this needs to be confirmed in the following for the specific fabrication scheme as outlined in Chapter 4. Two lines with the same arm lengths for single-arm antennas were also written into the same write-field. To evaluate fabrication errors, each line was fabricated twice. A representative set of two-arm antennas of arm lengths 50 nm through 25 nm as observed using SEM and AFM imaging is shown in Figure 6.1. In the SEM images, a slight 'halo' effect around the structures can be observed. This is an effect only visible starting with the second scan of a structure. This peculiarity is thus attributed to contaminants polymerized by the electron beam. Discussions at a Raith company product user meeting revealed that other users producing gold nanostructures of comparable sizes observe similar effects, but there is no definite explanation for the phenomenon.

Real Colour CCD Images

In a first step, the specific section was imaged using the real colour CCD camera to yield an overview of the response. These images are presented here to allow for a quick intuitive understanding of the antenna scattering properties. For the first image, the polarizer in the detection channel was set to only let pass the longitudinal scattering, i.e. along the long antenna axis. The image obtained for the set of antennas described above is given in Figure 6.2. A few observations can be made right away. First, two-arm antennas (on the left side) have a much stronger response compared to their single arm counterparts (shown on the right). Also, for both types, the scattering response strongly increases with size. Finally, the colour of the response seems to be fairly red, but there are indications of a green response for the smallest structures, especially for the single arm antennas. These preliminary findings will be confirmed and quantified using spectrally resolved measurements in the next section. Before

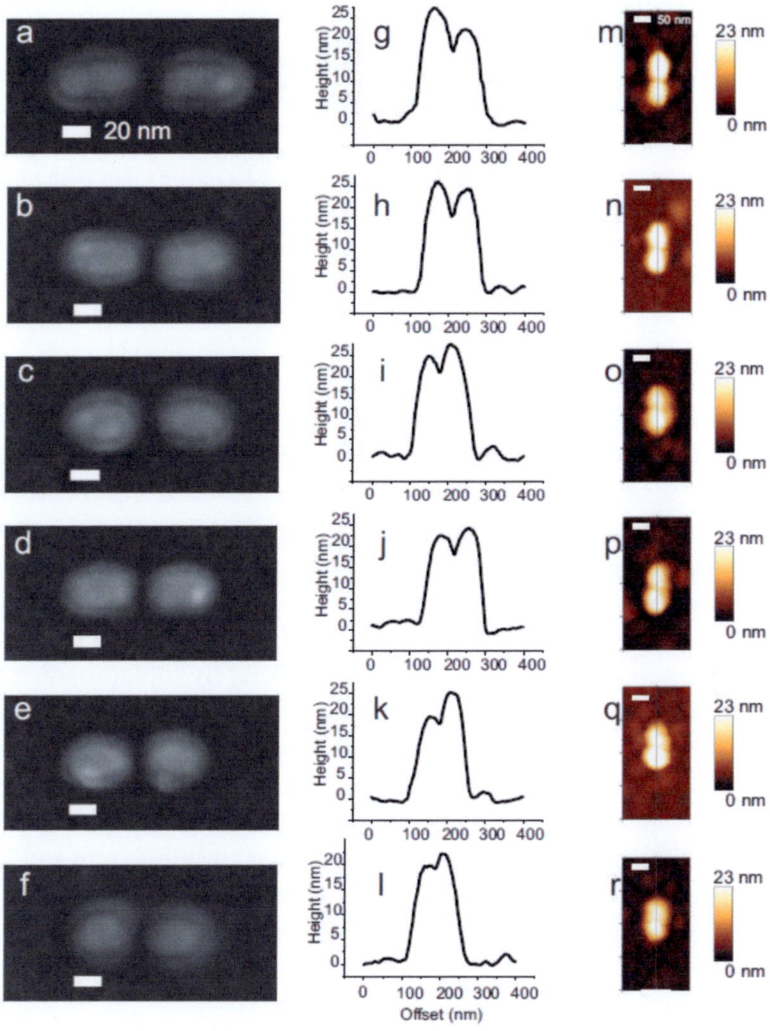

Figure 6.1: a)-f) SEM images of two-arm antennas of 50 nm through 25 nm arm length, reduced in 5 nm steps, g)-l) respective AFM cross-sections through the corresponding AFM images m)-r).

Figure 6.2: Real colour CCD image of the *longitudinal* scattering response of two-arm (left, nominal gap size 20 nm) and single-arm (right) gold nanoantennas with arm lengths of 25 nm through 65 nm (lower two lines) and 70 nm through 110 nm (upper two lines), with each arm varied in 5 nm steps.

continuing, it may be worthwhile to also take a look at the transverse scattering. As pointed out in the introductory sections, elliptical particles also have resonances along their short axis, and the same of course holds true for the antenna arms presented here. Obviously, the scattering response should be clearly shifted towards shorter wavelengths, as the width (in this case 20 nm) is much shorter than the antenna length. The response should also be much weaker due to the proximity to interband relaxation channels. Finally, antenna coupling should have less influence on the scattering properties, and the response intensity should mainly be influenced by the amount of material present, while the spectrum obtained should be fairly constant. A small aspect ratio dependence of the luminescence is also possible, but it is expected to be very small. (In fact, this will be confirmed by simulations in Chapter 7, where the two-photon luminescence response of such antennas is investigated.) All these preliminary assumptions are confirmed to a certain extent in Figure 6.3. In fact, the transverse scattering can be observed and is in the green spectral range, but is very weak and only seen under ideal measurement circumstances. For the smaller set of antennas from 25 nm to 65 nm, it is invisible for single arm structures. Both for single-arm as for two-arm antennas, it can be neglected in polarization-independent spectral measurements for those structures compared to the transverse response. In fact, it was impossible to use the spectral imaging to resolve transverse spectra of such structures (presumably due to lower efficiency of the EMCCD camera at lower wavelengths), which is why only the longitudinal modes will be investigated more thoroughly to obtain the spectrum in the following. However, additional details on transverse spectra can be found in Chapter 7 confirming that the spectra are indeed found in the green wavelength regime.

Figure 6.3: Real colour CCD image of the *transverse* scattering response of two-arm (left, nominal gap size 20 nm) and single-arm (right) gold nanoantennas with arm lengths of 25 nm through 65 nm (lower two lines) and 70 nm through 110 nm (upper two lines), with each arm varied in 5 nm steps.

Scattering Spectra

Fig. 6.4 depicts a scattering intensity spectrogram of several two-arm gold dipole nanoantennas of length $L_{arm} + \Delta + L_{arm}$, where Δ is the size of the antenna feed gap. Spatial positions of the individual antennas are retained on the y-axis, whereas the spectral scattering intensity information is encoded as false colours versus the linear photon energy scale between 1.5 eV and 2.5 eV on the x-axis.

This set gives as an example the total scattering intensity versus photon energy for the nine smallest two arm antennas with decreasing antenna arm length L_{arm} from top to bottom (from 65 nm to 25 nm in 5 nm steps) with constant feed gap size Δ and width W of 20 nm each. As expected from the real colour images, the gold nanoantenna resonance is found to be a function of L_{arm}. The spectral peaks are shifted towards higher energies for shorter structures. Also, the peak scattering intensity decreases for shorter antennas. The vertical stripes in the image in the lower energy regime are an artifact introduced by the EMCCD camera for detection at higher wavelengths (etaloning due to back-thinned EMCCD).

This is why no spectra could be taken for the longer antenna structures described above (L_{arm} = 70 nm and above), as their main resonances lie within this region. It is also obvious that the scattering intensity is mainly damped beyond 2.45 eV. As stated previously, this suppression of the scattering response can be attributed to a small interband transition of gold in this spectral range, where absorption and thus Ohmic losses are dominant. Any transverse features are also close to this absorption range and therefore generally weak, as we have already seen in the true colour CCD images. The skin depth of gold

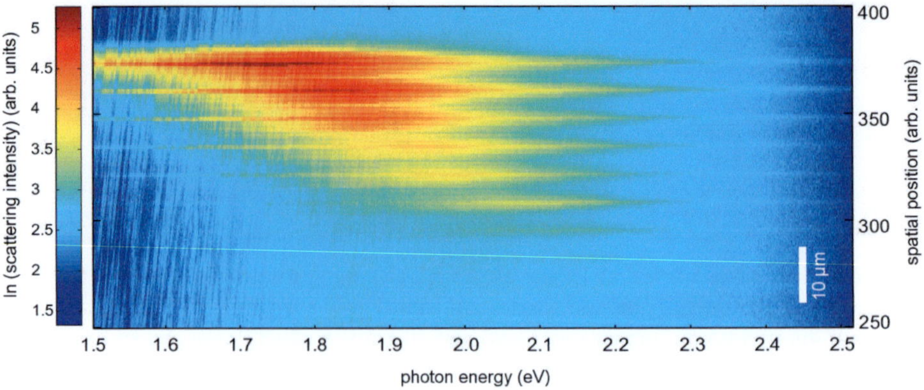

Figure 6.4: Scattering spectra (no polarization filter introduced) as obtained as a multi-dimensional spectrogram for two-arm antennas of constant gap size of 20 nm and varying arm length from 65 nm to 25 nm in 5 nm steps from top to bottom.

in this visible spectral range around 2.45 eV is approximately 20 nm, which is exactly on the order of the smallest feature size of the gold nanoantennas.

Figure 6.5 a) again depicts the scattering spectra of individual two arm gold nano-antennas from Figure 6.4 with antenna total sizes of $L_{arm} + \Delta + L_{arm}$ as indicated, a constant width of 20 nm, and height of 30 nm. This time, the spectral bins were integrated and corrected for the spectral properties of the excitation source. Spectra for corresponding single arm antennas of length $L_{single} = L_{arm}$ with the same height and width are shown in Figure 6.5 b). All curves are drawn to scale, i.e. while the scattering intensity units are arbitrary, the specific amount of scattering is decreasing significantly for the smallest structures, which is also why the tiniest structures are shown in a separate diagram as an inset.

As antennas can be treated as damped Lorentz oscillators, a Lorentzian squared fit procedure is used to analyse the spectral scattering peak position as a function of L_{arm}. With such a fit, the spectral width and the numerical integral of the scattering photon energy profile can also be obtained. Using the values for the integrated scattering energy and the peak scattering wavelengths, it can be shown that both antenna architectures show a general common trend that is summarised in Figure 6.6: With increasing L_{arm}, the resonance peak position shifts to lower energy values. This nicely confirms the scaling law derived from Fabry-Pérot resonances for single-arm antennas presented in Chapter 2.2.3. Both nanoantenna architectures studied here show a nearly linear

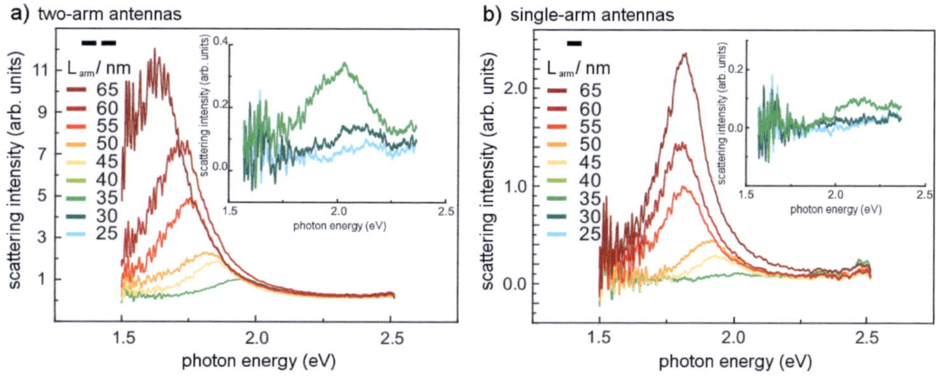

Figure 6.5: a) linear scattering intensity spectra of the longitudinal mode of two-arm nanoantennas of total length (L_{arm} + Δ + L_{arm}), with L_{arm} of 25 nm to 65 nm, gap size Δ = 20 nm; b) scattering spectra of the longitudinal mode of single-arm nanoantennas of total length ($L_{single} = L_{arm}$) of 65 nm to 25 nm.

energy dispersion as a function of L_{arm} within the photon energy range of 1.5 eV to 2.5 eV.

For the single-arm antennas, a direct comparison to the quantitative predictions from the Fabry-Pérot model by Novotny [137] can be made. The evaluation of Equation (2.59) for gold rods with a radius R of 10 nm (corresponding to the width of 20 nm), an effective refractive index of about 1.3 (considering an air-glass/ITO interface), $\epsilon_\infty = 9.84$, and $\hbar\omega_p = 9.03$ eV leads to the relationship shown in Figure 6.7 as a dashed line in comparison with the measured data (dots). The model is thus confirmed for gold single-arm antennas in the visible regime, albeit small differences are present, which may be due e.g. to the not perfectly cylindrical shape of the fabricated nanoantennas.

It has already been mentioned that the antennas of 20 nm width did not show any measurable transverse response on the EMCCD camera attached to the spectrometer used. However, it was possible to obtain spectra of the transverse mode for thicker antennas. To this purpose, the gold two-arm nanoantenna width W was increased to up to 70 nm while the total antenna length L_{arm} + Δ + L_{arm} was kept constant at 300 nm (a long antenna was chosen for this experiment to allow for the collection of a large number of photons). The spectral shift of the transverse mode as a function of antenna width can clearly be resolved. This is depicted in Figure 6.8. The spectra show that aspect-ratio considerations as described in Chapter 2.2.2 matter for gold nano-antennas, as the transverse spectrum for a width of 60 nm has a clearly shifted peak compared to the single-arm antenna of 60 nm length. Unfortunately, below a

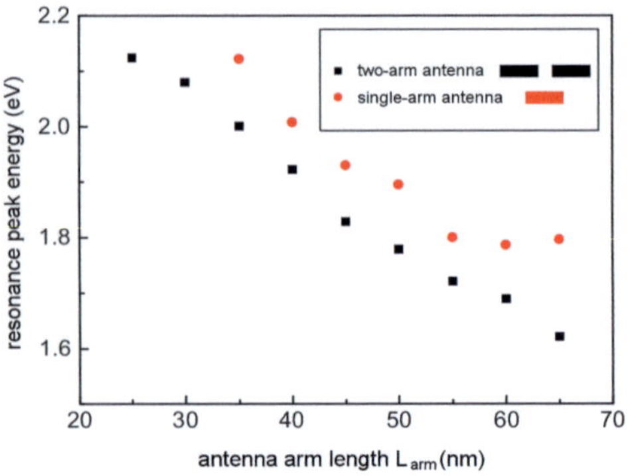

Figure 6.6: Longitudinal resonance scattering energy vs. antenna arm length L_{arm} for two-arm and single-arm gold dipole nanoantennas.

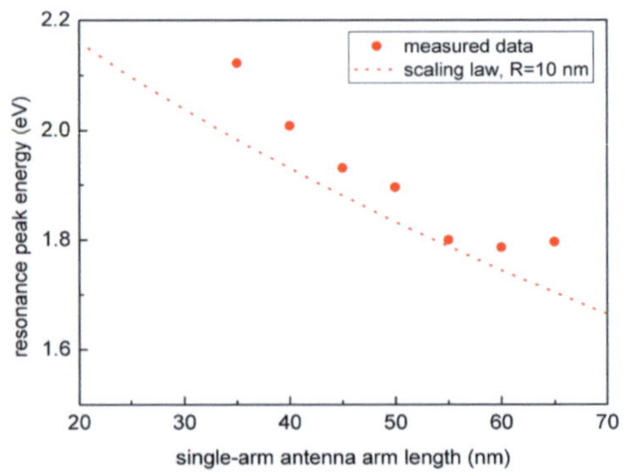

Figure 6.7: Comparison of the scaling law from Equation (2.59) (dashed line) to the experimentally measured single-arm gold antenna scattering peaks (dots). The slight deviations from linearity for the theoretical curve stem from the fact that it is originally linear in wavelength, so it is somewhat curved for a plot using an energy (and thus inverse) scale.

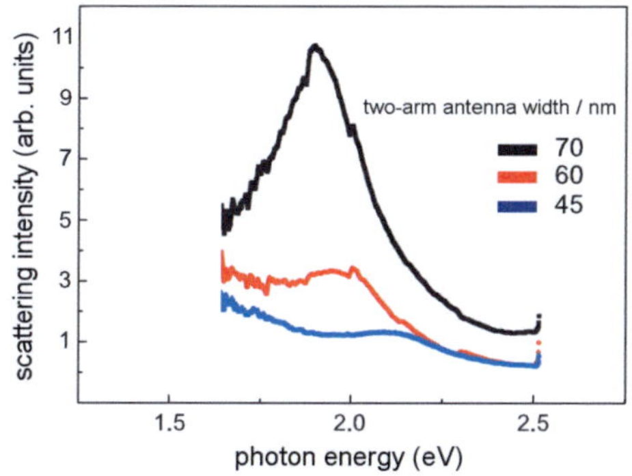

Figure 6.8: Transverse scattering as observed from gold two-arm antennas of arm length 140 nm with a 20 nm gap. Widths of 70 nm, 60 nm, and 45 nm were investigated, below that width, no transverse spectra could be obtained.

nanoantenna width of 40 nm any signature of a transverse mode in our white light scattering experiment for the signal-to-noise level available disappears, even though it was visible using the real colour CCD camera. This can be seen as advantage, as well, however, as then a polarization filter is not required in the detection channel to obtain the longitudinal spectra, which makes the antenna response more similar to classical RF responses.

6.1.2 Antenna Gap Variation

For studies on the effect of gap variation on the longitudinal resonance, the two-arm antenna gap size was varied in steps of 5 nm, from 20 nm to 35 nm for antennas with different arm lengths. Additionally, antennas with comparably large gaps of 60 nm and 700 nm were fabricated, which should (at least for the largest gap size) lead to a response of two uncoupled arms. Two questions were to be addressed with this study. First, the influence of the size of the antenna gap on the (peak) resonance frequency. Second, the influence on the amount of light scattered into the far-field. Figure 6.9, where the gap width Δ is varied from 60 nm down to 20 nm, with L_{arm} as a parameter, shows the influence of the gap size on the resonance peak. The scattering resonance spectrally red-shifts below a gap feature size of 30 nm. As already introduced in the theoretical section, it can be assumed that the introduction of the antenna gap as a design element of optical antenna architecture further reduces the Coulomb field due to the presence of an additional dipole field comprised of charges that pile up at the antenna gap. This then perturbs the internal field in

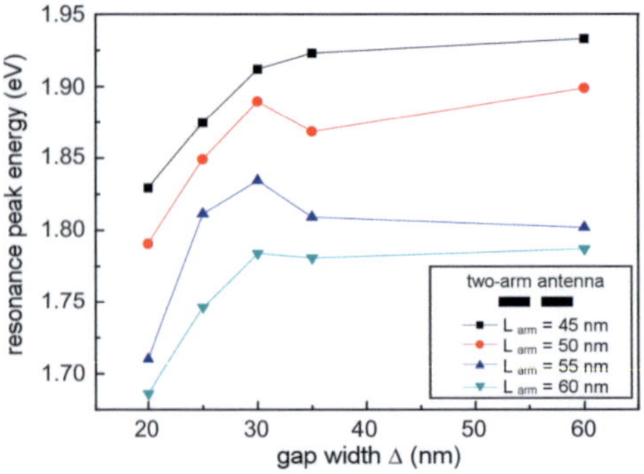

Figure 6.9: Shift of the peak scattering energy with antenna gap size Δ variation for two-arm antennas of different arm length L_{arm}.

the gold dipole nanoantenna with a gap compared to the single-arm antenna architecture, which then leads to the spectral shifts. Not visible in these far-field studies, but expected for near-field studies (as will be seen in a later chapter), the antenna gap may be seen as a designed electromagnetic hot spot where the localised field strength can be influenced tremendously by its feature size Δ. The fact that the scattering wavelength shifts towards the red can be seen as a slight disadvantage, as it in fact requires even smaller antenna arms to be fabricated to allow for higher peak scattering energies. On the other hand, once a specific arm length has been chosen, many lower energy peak resonances can be reached by the variation of the gap alone. This has specific advantages for antenna fabrication by chemical means, where often only a building block of one specific size can be generated. Of course, for this approach to be successful, an effective method for antenna arm positioning and influencing the gap would need to be found.

To also investigate the influence of the gap size on antenna scattering intensity, scattering spectra of several two-arm antennas with arms of constant length L_{arm} of 60 nm, but different gap sizes Δ (700 nm, 35 nm, 30 nm, 25 nm, and 20 nm) were also taken and compared to the corresponding spectra of a single-arm antenna of length $L_{single} = L_{arm}$. The resulting spectra, again all shown with an arbitrary, but intensity conserving scale, can be seen in Figure 6.10 a). As before, the antenna width of 20 nm was kept constant for all structures. The single-arm antenna shows the lowest scattering intensity, while the two-arm antenna with a 700 nm gap scatters about twice the amount of light with a very similar spectral position as the single-arm one. From the spectral shape of the scattering energy profile of the 700 nm gap nanoantenna (analysing the resonance peak position and the spectral width), it can be concluded that this gold antenna does not show any near-field coupling (it may experience some comparably weak far-field coupling) with a pronounced increase in radiation damping, since the far-field electromagnetic coupling of the two gold antenna arms may interfere constructively, leading to an enhanced radiation damping mechanism [44]. The fact that the spectrum of the 700 nm gap antenna shows two peaks in close proximity is most likely due to slight shape variations of the two arms (compare the previous chapter on fabrication), which can never be completely excluded.

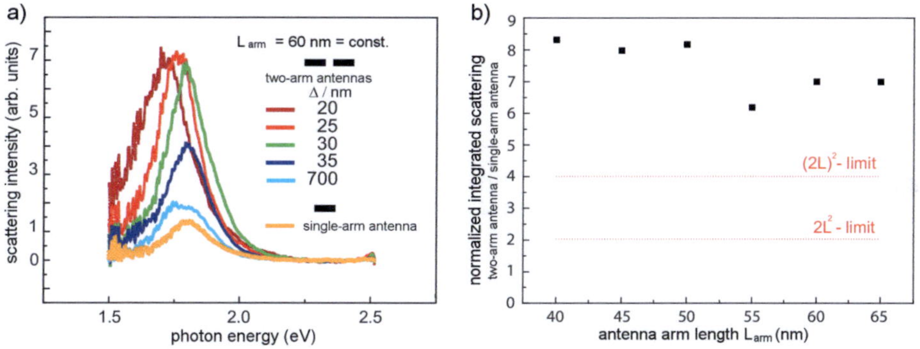

Figure 6.10: a) Spectral shape for several two-arm antennas with constant L_{arm} = 60 nm and varying gap size Δ, b) integrated intensity for two-arm antennas with Δ = 20 nm compared to the integrated intensity for single arm structures with $L_{single} = L_{arm}$.

For the small gap sizes Δ of 35 nm down to 20 nm (still with L_{arm} = 60 nm), apart from the red-shift in resonance peak frequency which has been discussed previously, the scattering intensity increases drastically. An enhancement of a factor of seven is obtained for the smallest gap sizes. This finding is not limited to this specific arm length of 60 nm, but is confirmed for additional antenna arm lengths in Fig. 6.10 b), where the integrated scattering intensity of several two arm antennas with a 20 nm gap with their respective single arm counterpart with corresponding $L_{single} = L_{arm}$ is compared. The increase in intensity is again by a factor of 7 to 8.

While some increase in scattered intensity is expected for two-arm antennas (even without involving any complicated physical effects, as just twice the amount of material is present), the measured intensity enhancement is surprisingly high when the simple V^2 proportionality of the scattering cross section is taken into account. While this proportionality has to be applied with care when comparing structures of different L_{arm} to each other (since total length and resonance wavelength change substantially), it is, however, a good estimate for the increase in the amount of light scattered by the two-arm antenna with arms of length L_{arm} each compared to their $L_{single} = L_{arm}$ counterpart – as had been seen earlier, the shift in resonance frequency for those corresponding antennas is comparably small (only about 0.1 eV). It is interesting to see that the scattering enhancement is stronger than expected compared to either the scattering cross section scaling with a factor of 2 (for simply considering the same structure twice, as for structures separated by a large gap) or a factor of 4 (for taking into account the total length of two arms coupled by an infinitesimally small gap, i.e. $L_{arm} + \Delta + L_{arm}$, with $\Delta \rightarrow 0$). Further experiments combined with electromagnetic simulations are required to give more substantial insights into this observation. From the reproducibility of the results for different arm lengths, we may assume that the surface roughness and its contribution to the photon scattering linewidth is constant for the antenna arms of given aspect ratio L_{arm}/W (where the aspect ratio is defined for the single antenna arm, not for the structure with the gap). Thus the designed antenna feed gap Δ apparently not only has a significant function as the coupling near-field unit of two antenna arms in order to localise electromagnetic fields at sub-wavelength volume. Moreover, this gap also seems to mediate an exceptionally enhanced far-field scattering probability of photons that had not been observed or discussed before this study, as spectra were typically shown individually normalized to unity. On a side note, this increase in scattering can also not be explained exclusively by a more directional emission of the dipole structure – if a change from an omnidirectional Hertzian dipole emitter to an ideal half-wavelength dipole antenna is considered, this would increase the collected scattering intensity by about 10% at best. The enhanced fields localized at the gap therefore need to be mediators for the observed increase in linear scattering. Additional experiments are required to give more insight into the near-field contribution (and thus potentially the far-field influence) of those designed hot spots in two-arm nanoantennas. A first step into that direction will be shown in Chapter 8.

6.2 Cross-Resonant Four-Arm Gold Antennas

It was already briefly mentioned in the theoretical section that a combination of two different two-arm antennas oriented with a rotation of 90 degrees and a common antenna gap could lead to interesting effects. While they could be used to efficiently couple to circularly polarized light [142, 19], the interest in this short section shall be how efficiently they can be used to separate (by polarization) two different main resonances. One could think for example of a fluorescent structure like a quantum dot that is placed into such a cross structure. The fluorescence generated by the quantum dot would then lose the polarization properties from the excitation light (as fluorescence is generated from electronic states), and at least 50% of the emission would then be enhanced by the smaller antenna making available radiative channels for the blue-shifted emission. Of course, there are several obstacles to overcome to achieve such a combined structure (e.g. the quantum dot has to be placed in the proper position, and the modifications to the resonance by the introduction of the quantum dot have to be taken into account). Thus, while the coupling of the single quantum dot to a nanoantenna is beyond the scope of this thesis (successful placement of quantum dots or diamond defect structures close to optical antennas has for example been shown very recently by Curto et al. [42] and Schell et al. [159]), it is nonetheless interesting to investigate four-arm structures for their spectral properties, having this possible application in mind.

Long arms and short arms were thus combined for four-arm structures, again using the previously outlined electron-beam lithography protocol. Due to cross-exposure effects between the arms, these already tended to form connections between separate arms for comparably large gap sizes. Gap sizes of less than 30 nm between two opposing arms can at the moment not be fabricated reproducibly for gold four-arm nanoantennas. This is unfortunate, as coupling only sets in at about this value, as was shown in Figure 6.9. On the other hand, the gap between two adjacent arms can be very small and be less than 10 nm. This implies that coupling is very likely to occur between those adjacent arms, which needs to be taken into account for the design of the antenna.

Two exemplary four-arm antennas with a nominal short arm length of 40 nm each and long arm lengths of 55 nm and 95 nm, respectively, and a 45 nm gap, can be found in Figure 6.11. Dark-field measurements of the resonances along the long and short antenna axes (data not shown) for several such structures with a further variation in short and long arm length only reveal that the scattering peaks are very similar to the peaks of single arm structures, so coupling is unfortunately not present, which also makes the presence of strongly enhanced fields in the gap as required for the above-mentioned application unlikely. More details on the investigation of such structures can be found in the *Studienarbeit* by Patrick Schwab [165].

Summarizing, the fabrication of efficiently coupling four-arm structures made by combining two coupled two-arm antennas using electron-beam litho-

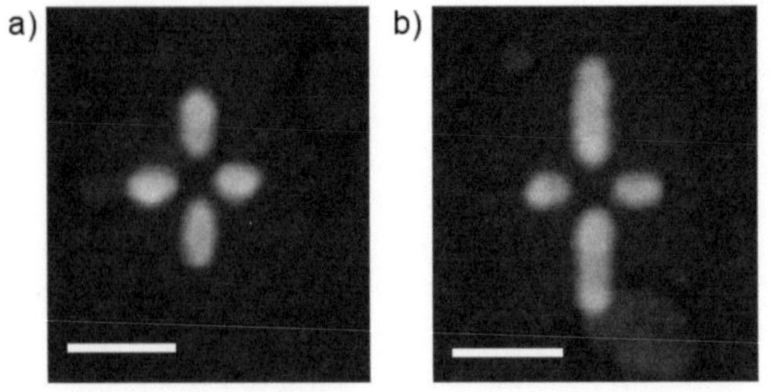

Figure 6.11: Exemplary four-arm gold antenna structures as imaged using an SEM with nominal short arm length of 40 nm each and long arm lengths of a) 55 nm and b) 95 nm. Scale bar length is 100 nm. (Images courtesy of Patrick Schwab)

graphy alone is challenging, even though it might be possible with elaborate single dot exposure schemes or through the use of other resists allowing for even smaller linewidths. It might also be possible to use other materials, where larger gaps might be allowed to still enable coupling. One other material that was studied for single-arm and two-arm antennas is aluminum. General trends on aluminum structures will thus be presented in the next section.

6.3 Aluminum Antenna Scattering

In Chapter 2, it was already seen that the material properties of aluminum greatly differ from gold, most notably including an additional interband transition around 1.75 eV. Also, aluminum tends to oxidize rapidly in air, which yields a thin layer of aluminum oxide (typical thickness of 4 nm) at the exposed surface [145]. Scattering spectra of aluminum nanoantennas will yield very different properties, as the dielectric function is very important in determining nanostructure resonance. For a comparison, aluminum antennas with similar feature sizes as the gold structures were investigated. It should be noted that there is some small difference in fabrication scheme, as the aluminum antennas were fabricated directly on the glass substrate (compare Chapter 4) to allow for additional experiments at a later stage beyond the scope of this thesis (e.g. the tip-assisted movement of quantum dots on the surface, which is easier on a smooth glass substrate than on an ITO layer).

There are thus small shifts in resonance behaviour that are due to the slight change in dielectric environment of the antennas. However, for comparison

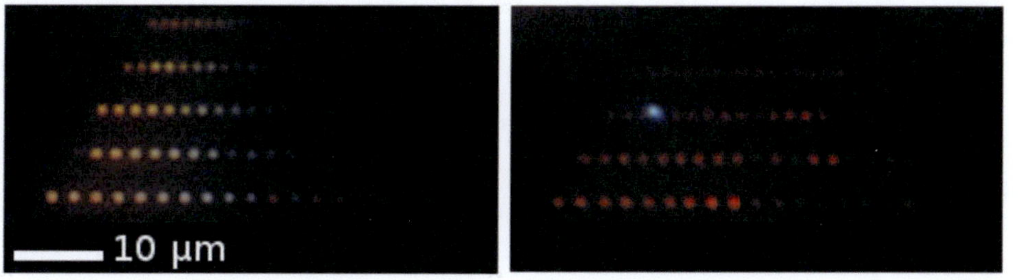

Figure 6.12: Real colour CCD image of the *longitudinal* scattering response of two-arm nanoantennas with various arm lengths (left to right) and width of 30 nm through 50 nm (top to down) for aluminum (left) and gold (right), with all sizes varied in 5 nm steps.

a batch of aluminum antennas was also fabricated on an ITO layer, and the spectral shifts were found to be negligible. Compared to the investigations with gold nanoantennas, the width for all structures is now 30 nm, as a width of 20 nm proved too small to allow for an acceptable SNR in the detection channel.

To again allow for a quick and colorful introduction to the aluminum antenna scattering behaviour, Figure 6.12 gives a dark-field image for a set of structures of the same size for aluminum (left) and gold (right). Directly visible is that gold antenna longitudinal resonance of appreciable scattering intensity is almost exclusively available in the red spectral regime, while the aluminum scattering response easily encompasses the green part of the spectrum and even shows some peaks in the blue. This blue emission is again more challenging to resolve spectrally due to more limited detection capabilities at those frequencies for the spectroscopic devices available. It should also be noted that while the smallest gold structures of 25 nm arm length still yield a significant response with an acceptable SNR, the aluminum antenna structures with a length of less than 70 nm can hardly be resolved.

6.3.1 Aluminum Antenna Arm Length Variation

For aluminum antennas, similar images as the ones presented in the first section of this chapter for gold can be presented. CCD real colour images again give valuable first information on the scattering properties. The longitudinal mode for two-arm and single-arm antennas is presented in Figure 6.13 for aluminum antennas of constant width of 30 nm and varying arm length (this time from 30 nm to 200 nm in 10 nm steps, with a 20 nm gap for the two-arm struc-

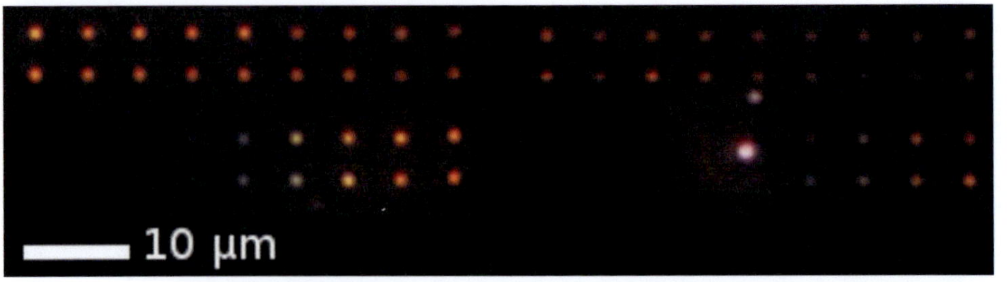

Figure 6.13: Real colour CCD image of the *longitudinal* scattering response of two-arm (left, nominal gap size 20 nm) and single-arm (right) aluminum nanoantennas with arm lengths of 30 nm through 110 nm (lower two lines) and 120 nm through 200 nm (upper two lines), with each arm varied in 10 nm steps.

tures). Again two-arm structure resonances seem to be shifted slightly towards higher resonance wavelengths for the same arm length compared to single-arm structures, but a spectroscopical analysis is once more required to learn more about this behaviour.

Likewise, the transverse mode under length variation can be examined. The scattered signal is very weak, however, so no image can be shown. This is not surprising considering that the longitudinal resonances seem to be shifted towards higher energies for similar arm lengths compared to gold, which implies that the transverse resonance is most likely to occur in the UV spectral range.

Spectral Investigation

A spectral investigation of the scattering behaviour of the aluminum antennas was also performed as previously for the gold antennas. As only a fewer amount of structures could be placed in a single antenna field due to crosstalk from the edges of the glass window within the niobium film, excitation conditions for two-arm and single-arm structures are not perfectly the same for the scattering measurement of the two types (as they were located in different fields), such that only normalized spectra are presented in the following. Longitudinal spectra for aluminum two-arm antennas (20 nm gap) and single-arm antennas with arm lengths from 95 nm through 200 nm, with 30 nm width, are shown in Figure 6.14. As already expected from the real colour CCD images, generally both types of antennas yield again a blue-shifted response with a reduction in arm length. For smaller structures, spectra can also achieve peak wavelengths in the green, which was hardly possible for gold nanoantennas.

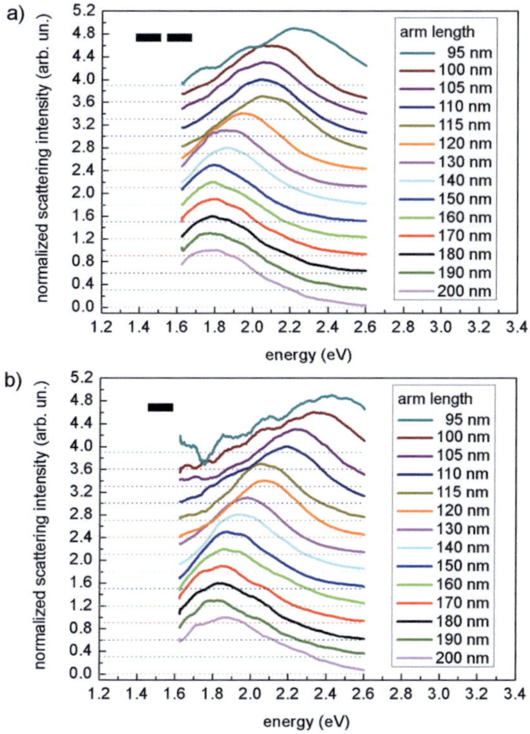

Figure 6.14: Scattering spectra for a) two-arm and b) single-arm aluminum nanoantennas of 30 nm width, a 20 nm gap (for two-arm structures), and arm lengths as indicated. All spectra have been normalized to unity and and offset of 0.3 has been introduced between individual spectra to facilitate discrimination between them.

Also, for the same resonance energy, the aluminum antenna clearly needs to be longer compared to gold, thus giving a total length closer to the $\frac{\lambda}{2}$ resonance of radiofrequency antennas. This is in good agreement with the fact that aluminum has a smaller skin-depth than gold at optical frequencies (compare Table 2.2). Also in good agreement with theory (see e.g. reference [137]) is the fact that interband effects become visible for aluminum towards the red end of the visible spectrum (around 1.75 eV), where suddenly the resonance peaks both for two-arm and single-arm antennas stop to shift and remain approximately constant.

Despite the presence of the aluminum-oxide layer, two-arm antennas still couple and are thus red-shifted compared to the single-arm counterparts, as again shown in a view of the peak wavelengths for single-arm and two-arm aluminum nanoantennas in Figure 6.15. Somewhat surprisingly, the spectra

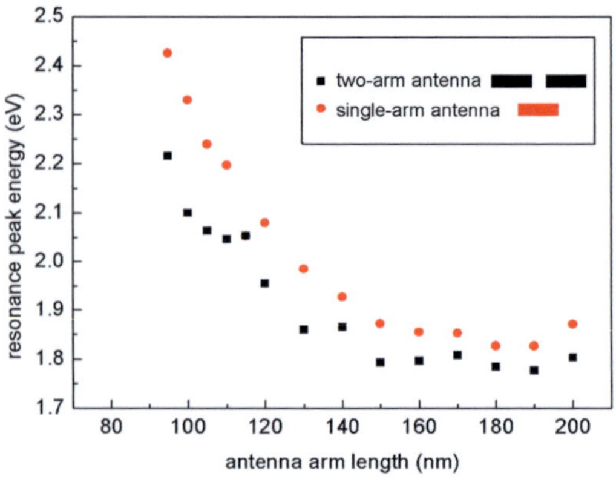

Figure 6.15: Scattering peak energies for aluminum two-arm and single-arm antennas. Two-arm antenna scattering peaks of the same arm length are red-shifted compared to single-arm ones. The interband transition around 1.75 eV causes the peak energies to remain constant for antenna arm lengths between 150 nm and 200 nm.

even differ when the interband transition region is reached, as two-arm antennas also settle at a lower energy for larger arm lengths.

Again a comparison can be made to the numerical model by Novotny (for the peaks outside the interband transition), although the parameters for the rod width and the refractive index are more difficult to estimate than for gold due to the formation of the oxide layer. With a typical oxide layer thickness of about 4 nm, one would assume that the model can be used for $R = \frac{30-2(4)}{2}$ nm = 11 nm, but in fact a better agreement is found by assuming $R = 9$ nm, as seen in Figure 6.16. Other parameters used with Equation (2.59) are $\epsilon_\infty = 0.2$, $\hbar\omega_\mathrm{p} =$ 14.76 eV, and an effective refractive index of 1.5 (increased in comparison to the gold antennas due to the contribution of the aluminum oxide). This time, for a good agreement to the model a smaller radius than presumably present has to be used, possibly due to more complicated oxidization effects or a lesser validity of the model due to the presence of the interband transitions, which cannot be explained by simple linear scaling with wavelength.

The effect of a variation in antenna width for a constant arm length on the longitudinal scattering peak was also investigated. For a constant arm length between 100 nm and 150 nm, no noticeable shift in peak wavelength could be observed when the structure width was increased from 20 nm over 30 nm to 50 nm (spectra not shown here). This may partly be due to the greater difference between length and width in general for aluminum structures, which makes the change in aspect ration less significant compared to gold structures,

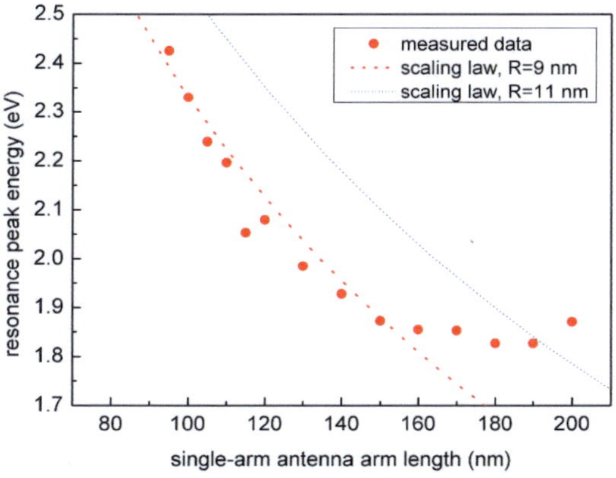

Figure 6.16: Comparison of the scaling law from Equation (2.59) for two different radii R (dashed lines) to the experimentally measured single-arm aluminum antenna scattering peaks (dots). The slight deviations from linearity for the theoretical curve stem from the fact that it is originally linear in wavelength, so it is somewhat curved for a plot using an energy (and thus inverse) scale.

where such aspect ration influences have been observed [72]. The absence of a shift due to aspect ratio variations is, on the other hand, in strong contradiction to the scaling law just applied, as Equation (2.59) contains a strong radius and thus aspect ratio dependence. A small blue-shift for the thinnest antennas cannot be ruled out completely. More details on this topic can be found in the *Diplomarbeit* by Patrick Schwab [166].

6.4 Comparison of Gold and Aluminum Antennas

A comparison of 30 nm wide aluminum structures and 20 nm wide gold structures in terms of their spectral behaviour when only the length is varied is carried out in Figure 6.17 a), where peak resonance energies for gold and aluminum two-arm and single-arm structures are depicted. While the exact peak wavelengths may not be comparable due to the difference in substrate used (gold on glass/ITO vs. aluminum solely on glass) and the fact that the aluminum structures were slightly wider (30 nm instead of 20 nm), some general trends can be directly seen and are summarized in the following: i) aluminum resonances are strongly blue-shifted for the same structure size, ii) aluminum structures are favourable for the green (and possibly the blue) spectral regime, while gold is better to be used in the red spectral range, iii) interband transitions become apparent for aluminum around 1.75 eV as expected from theory (compare Figure 2.7), and iv) the coupling behaviour both for gold and aluminum antennas is very similar, in both cases the coupled structures have resonances shifted towards the red, which is in good agreement with theory. It is noteworthy that this red-shift is somewhat stronger for aluminum (about 0.15 eV compared to 0.1 eV for gold). This might be due to influences of aluminum oxide on the surface, which has a higher refractive index than air or glass (about 1.76 [145]) and can thus be responsible for some additional shift if the fields are concentrated there.

A comparison of the scattering spectra of gold and aluminum also reveals a strongly increased spectral width (FWHM) for aluminum compared to gold (about 0.5 eV for the aluminum structures versus about 0.25 eV for the gold nanoantennas), as plotted in Figure 6.17 b). The quality factors Q (compare Chapter 2.2.4) obtained from the peak energy values and the corresponding FWHM are depicted in Figure 6.17 c). For both antenna materials, the FWHM is larger and thus the Q factor lower for two-arm antennas. This implies, as already discussed in Chapter 2.2.4, enhanced radiative or Ohmic losses. Since for gold antennas a strongly enhanced scattering emission intensity has been observed (compare Figure 6.6), at least part of the additional losses need to be radiative. Generally, aluminum has the greater spectral width and thus a lower Q factor, which is most likely mainly related to increased Ohmic losses, as the emission intensity is on about the same level as for gold structures at similar excitation intensities, albeit the aluminum structure sizes involved are much larger.

6.5 Conclusions

This chapter has depicted gold and aluminum nanoantenna scattering as observed via dark-field microscopy. With the nanoengineering e-beam lithography fabrication protocol presented in Chapter 4, two-arm and single-arm gold and aluminum dipole nano-antennas with feature sizes as small as 20 nm were reproducibly fabricated. For the smallest gold antennas, the linear optical scattering response function by systematic variation of the antenna dimensions was observed. The resonance energy peak position, the spectral width and the relative scattering intensity were determined. For gold structures, a nanoantenna feed gap with Δ smaller 30 nm synergistically combines the near-field coupling of two antenna arms of length L_{arm} each, creating localised electromagnetic hot spots at a well-chosen sub-wavelength volume, with exception-

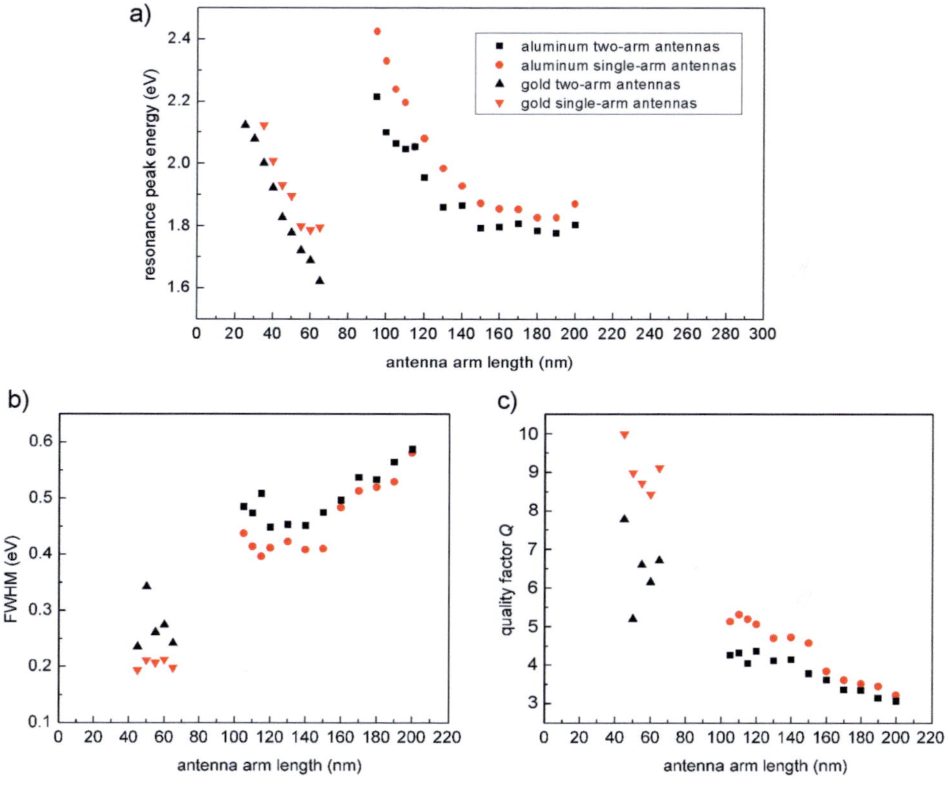

Figure 6.17: a) scattering peak energies for gold and aluminum two-arm and single-arm antennas, b) FWHM for the larger antenna structures, c) quality factors Q as derived from a) and b). More details can be found in the text.

ally enhanced far-field photon scattering probabilities. The physical volume does not explain the enhanced photon scattering probability completely when classical scattering theory is applied. It can be speculated that the nanoantenna feed gap as an electromagnetic hot spot acts as an additional dipole source that highly perturbs the internal field of the gold nanoantenna and effectively contributes to the enhanced far-field scattering intensity of visible photons. For aluminum antennas, below the interband transition region, antenna resonances also scale with length, albeit the resonances occur at significantly higher energies for structures of the same size as gold nanoantennas. Again in comparison to gold, the spectral width is greatly increased, which can be seen as an advantage if a broadband antenna operating in the visible is desired. This may be the case e.g. for plasmonically enhanced solar cells or light emitting devices, as a broad spectral range needs to be covered by such devices.

Chapter 7

Two-Photon Luminescence

In the previous chapter, experiments to access the linear scattering response of gold and aluminum optical antennas were outlined. With the spectral information gathered there, it is now interesting to perform a different kind of investigation: to excite antennas at a specific frequency and with a specific polarization and check for the response received depending on the antenna size and shape, which can be performed using a laser. Such experiments are particularly interesting if a nonlinear response can be observed, as the excitation wavelength needs to be filtered out before detection. Two-photon induced luminescence (TPL) is one of the effects which has been observed previously for optical antennas [71, 128, 162]. However, in all those studies, great emphasis was put on the emission intensity of such structures, while the emission spectra of two-arm or single-arm antennas had not been investigated in a systematic way. There was some evidence that particle plasmon generation and far-field decay (and thus a spectral response similar to the scattering response) could be triggered by an energy transfer to nanostructures using a laser. This emission derived from high energy excitation [126] is typically attributed to excited d-band holes which combine with sp-band-electrons, finally enabling the emission of a particle plasmon [51], as already depicted in Figure 2.11 in the theoretical section.

Recently, it has been shown for single gold nanoparticles that such a plasmon generation process can also be triggered by two-photon excitation [25, 23], which is very similar to how two-photon processes in typical fluorescence processes (see also Chapter 3) are usually considered [103]. In the following, luminescence emission spectra for gold and aluminum two-arm antennas and single-arm antennas under two-photon excitation with a fixed pulsed laser excitation at a wavelength of 810 nm are presented. It will not only be shown that these spectra look very similar to the ones obtained using linear spectroscopy as described in the previous chapter, but also that the red-shift and significant emission intensity enhancement for two-arm antennas compared to their respective single-arm counterparts can also be observed under two-photon excitation for gold, while additional effects arise for the interband transition region of aluminum. Additionally, cross-polarized luminescence with respect to the excitation light can be observed for gold antennas. The main findings for gold antennas presented in this chapter have been published in the journals *Nano Letters* [189] and *Optics Express* [191].

7.1 Experimental

The antenna structures investigated are the same (from a newly produced sample) as the ones used in the previous chapter. For the optical experiments, the confocal microscopy characterization setup already detailed in Chapter 5.3 is used. Details on the laser and the filters used within the setup can be found in the caption of Figure 7.1.

To avoid destruction of the optical antennas under investigation, the power used for excitation needed to be carefully tuned. The typical laser power applied (measured in front of the objective lens) was 10 μW for resonant two-arm antenna structures. For structures far from the resonance condition, up to 1 mW power could be used. Each pulse of approximately 500 fs width (at a pulse repetition rate of 76 MHz, corrected for broadening, e.g. due to beam-expanding lenses) thus reached a maximum laser fluence I of about $1 \cdot 10^2 \frac{J}{m^2}$, assuming a focal area of π x 200^2 nm^2 and an objective lens transmission efficiency of 0.7 [1] at 810 nm wavelength. To excite single-arm antennas with a similar SNR, around 10 times the power required for the corresponding two-arm structures had to be used.

7.2 Gold Antenna Longitudinal Plasmon Response

7.2.1 Results and Discussion

Antenna emission intensity investigation

As it could be seen in the previous chapter on scattering experiments and as it has also already been described in the theoretical section, nanoantennas with appropriately tuned arm length can provide efficient coupling to the excitation

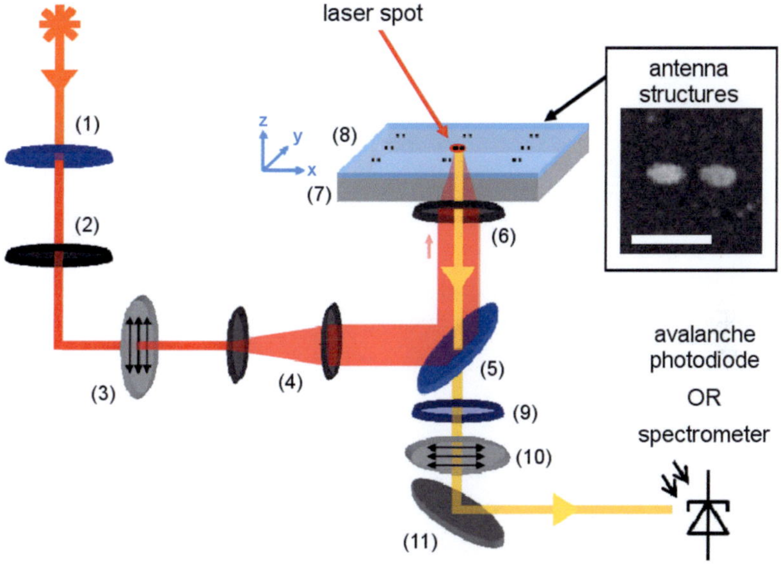

Figure 7.1: Confocal measurement setup for the antenna TPL plasmon response. The excitation laser (*Coherent Mira* Ti:Sa, 120 fs pulses, 76 MHz repetition rate, with center peak wavelength tuned to 810 nm) is first limited to wavelengths greater 775 nm by a long-pass filter (Semrock Edge Basic BLP01-785R-25, (1)). After optical density filters (2) and a polarizer (3), the beam is expanded (4) and sent, via a 50/50 beam-splitter (5) and a 100x-oil immersion objective (Zeiss, NA 1.46, (6)) onto the sample (7) with the nanoantennas (8). The emitted light is then again collected by the same objective lens. Only light with wavelength below 775 nm then passes through a short-pass filter (Semrock RazorEdge 785RU-25, (9)) in the detection path. Optionally, a polarizer can be introduced as well to allow for polarization sensitive spectral investigations (10). For detection, either an avalanche photodiode (APD) detector (for intensity investigations) or a spectrometer (Acton 2500i) equipped with an Andor EMCCD camera as a detection device, can be used, both accessible via a turnable mirror (11). The inset shows an SEM image of an antenna of nominal arm length 45 nm and gap width 20 nm, the scale bar length is 100 nm.

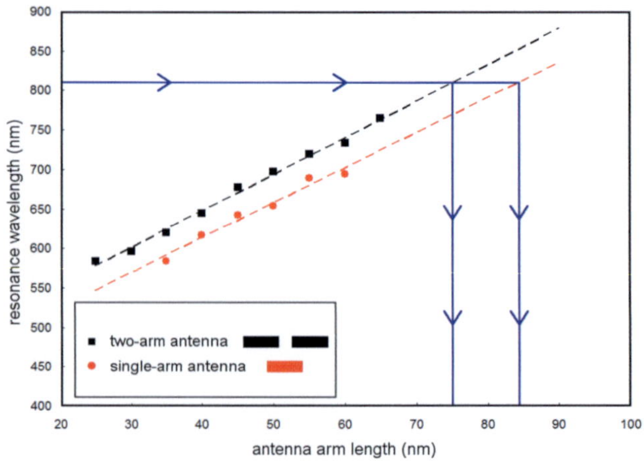

Figure 7.2: Extrapolation of the scattering results can be used to predict the resonant antenna arm length for a laser excitation wavelength of 810 nm. It is 75 nm for two-arm structures and 84 nm for single-arm ones.

light. From linear extrapolation of the scattering results presented, which in Chapter 2.2.3 has been shown to be appropriate, an arm length value for the strongest resonance at the chosen fixed excitation wavelength of 810 nm can be predicted. This is depicted in Figure 7.2. Thus a gold two-arm antenna is resonant for an arm length of about 75 nm, while about 84 nm are required for single-arm antennas.

If it is now assumed that the amount of emitted intensity directly depends on the amount of energy coupled into the nanostructure due to the fulfillment or non-fulfillment of the resonance condition, two-arm antennas with arm lengths below or above the resonant length will yield an emission intensity response that strongly decreases for larger length deviations from the resonant structure. As an immediate consequence, it is therefore challenging to excite many structures of different size with exactly the same power, as some may show no response at all for low irradiance powers while others already undergo irreversible shape transformations for higher excitation intensities. Up to now, there is no experimental data on the absolute two-photon absorption cross section for such gold nanoantennas. However, if the cross section for the two-photon mediated response in fact scales approximately like the linear scattering cross section, one can assume that there is almost no response if the excitation laser light is polarized along the short antenna axis of constant width 20 nm, as the nanoantenna is then severely detuned.

For a first experiment to check the influence of the polarization state and thus the efficiency of the coupling of the excitation light to the structure due to resonance conditions, an area including several antennas was raster-scanned, and the emitted intensity was collected using two different polarization states. Even though the predicted antenna resonance is for longer antenna arms and for reasons that will become clear once the spectra of the response are considered, the scan area only included two-arm antennas with arm lengths of 25 nm to 65 nm (the ones shown to yield a response in the visible in the previous chapter). The incident light was polarized either along the long antenna axis, Figure 7.3 a), or perpendicular to the long axis, Figure 7.3 b), while excitation wavelength and excitation power were kept constant. Evidently, there is no observable antenna response in Figure 7.3 b), while a response from the larger two-arm antennas can easily be detected in Figure 7.3 a). This observation is in agreement with expectations, and also with previous polarization dependent TPL measurements on single gold nanorods [23]. Also, for the polarization state along the long two-arm antenna axis, a first indication for the process that leads to light emission for that case is obtained: As not all structures generate an optical response for the excitation power chosen, it may be concluded that the excitation channel indeed relies on the outermost and decreasing (for shorter arm lengths) small absorption probability of the linear optical scattering response function. This is in strong contrast to nonlinear but non-resonant effects that can be observed on longer rods, which will be described in detail in Chapter 9.

As already described earlier, such emission intensity figures in which many antennas of different sizes are depicted always have to balance the excitation power to make sure that it is low enough to avoid destruction and structural deformations (e.g. from thermal heating [112, 200]) or detector saturation for highly resonant structures. At the same time, as many not perfectly resonant structures as possible should give a response discernible from the background signal. Figure 7.3 a) was obtained in that way, which allows to put together a plot for the total integrated intensity observed for each of the two-arm antennas. The total number of photon counts generated on the detector by each antenna were obtained by summing up all pixels from the raster-scan which included an antenna response and subtracting the average background value. The total photon counts for the antennas from Figure 7.3 a) are given in 7.3 c). For this plot, the polarization filter in the detection path was oriented such that only longitudinal components (i.e. electromagnetic waves polarized along the long antenna axis) could be detected. Obviously, the integrated longitu-

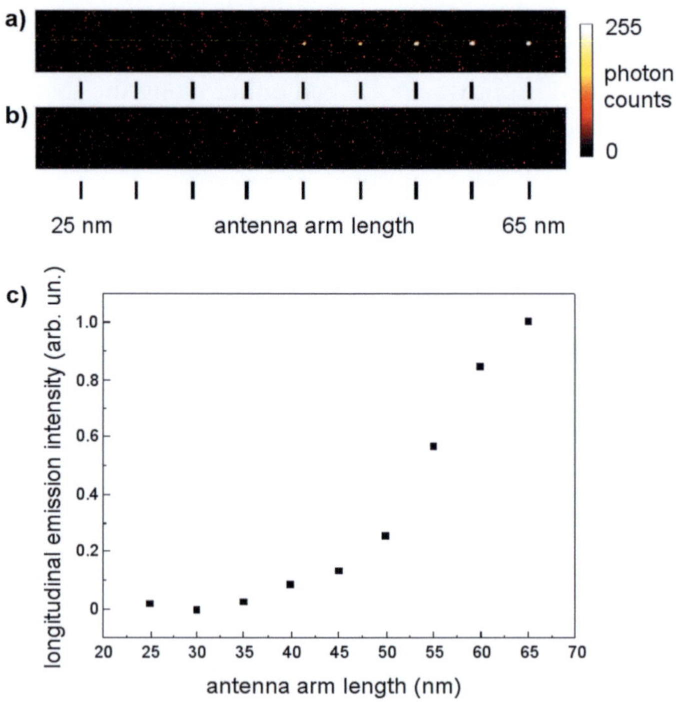

Figure 7.3: Optical intensity response (integrated intensity for λ<775 nm) for a scan field containing a set of nine two-arm antennas. a) gives the photon counts with the polarization of the incoming light tuned parallel to the long antenna axis, b) for the same nine antennas with the laser light polarized perpendicular to this axis. The distance between two neighbouring structures is 5 μm. c) Longitudinal excitation and photon emission intensity for two-arm antennas of different arm length with a fixed pulsed laser excitation intensity and wavelength (810 nm).

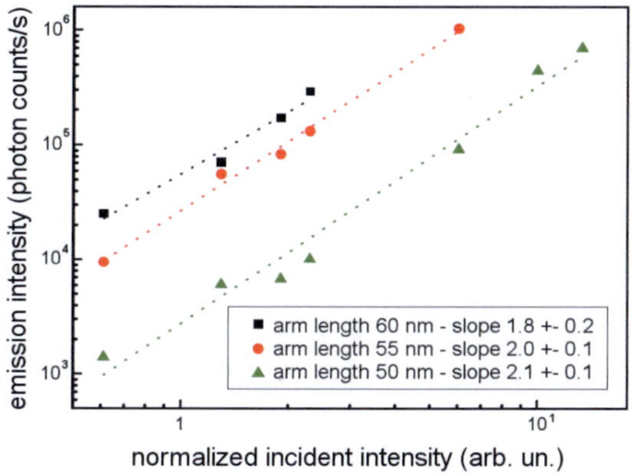

Figure 7.4: Double logarithmic plot of plasmon response intensity versus excitation intensity for gold two-arm antennas of 60 nm, 55 nm, and 50 nm arm length. The fitting curve leads to a slope of about 2 for all structures under investigation.

dinal antenna response intensity increases significantly when the arm length increases towards the resonant arm length known from the optical linear scattering experiments presented in the previous chapter. As already noted above, this is a first, but strong indication that the two-photon effect observed here is linked to antenna scattering properties in the excitation channel. It will be shown in the next section that the emission spectra are also strongly shaped by the particle plasmon. This is why the emission intensity of the perfectly resonant antenna as well as larger ones is not shown in Figure 7.3 c) – their longitudinal resonance plasmon peak is outside the spectral window available due to the filter that has to be inserted to eliminate the excitation light before detection.

The development in luminescence intensity response versus the excitation intensity for several two-arm antennas was also examined, as shown in Figure 7.4. The slope measured in the double-logarithmic curves is around 2 for all structures investigated, which indicates that mainly a two-photon luminescence signal is observed. Measurements for single-arm antennas yield the same slope (see also [23]).

When investigating the gold antenna scattering signal in the previous chapter, it was observed that the emitted far-field intensity from two-arm antennas was about a factor of eight higher than from corresponding single-arm antennas. For the nonlinearly generated emission light observed here, a similar comparison can be performed. For typical, non-destructive excitation intensities, a factor 10 to 65 higher photon count ratios for the two-arm structures compared to the single-arm ones were measured for constant excitation conditions. As the process is nonlinear, the exact amount of enhancement depends on the excitation intensity and thus also on the focus quality, which can change over time due to the immersion oil used to connect sample and objective. Nonetheless, independent of the exact amount of emission enhancement that can be achieved, the coupling mediated by the antenna gap plays a very important role in two-photon antenna response intensity. This claim, as already outlined in the theoretical chapters of this thesis, is also supported by numerical simulations which yield a strong and highly localized electro-magnetic field at the position of the antenna gap for small gap sizes at antenna resonance [156, 169]. The direct influence of the resonance condition on the amount of emitted two-photon luminescence for uncoupled structures had already been shown in a study of gold nanowires of different width and height by Kim et al. [90] and is supported by a simple theoretical model based on surface plasmon and lightning rod effects [34, 26]. With this measurement, additional evidence is provided that antenna coupling is also an important factor to be considered. An interesting expansion of this study would be to correlate variations in antenna gap size for structures of the same arm length (similar to what is presented for scattering in the previous chapter). If at some point additional fabrication techniques are available, it would be specifically interesting to investigate gap sizes even smaller than 5 nm.

TPL spectral investigation

After the investigation of the emission intensity, now a closer look at the spectra emitted shall be taken. It was already hinted in the previous section that the emission is directly correlated to the particle plasmon, so the luminescence will subsequently be termed plasmon response, which can be easily understood when looking at the spectra. Plasmon response spectra of individual two-arm gold antennas (total length L_{arm} + 20 nm gap + L_{arm}) and also single-arm antennas (length L_{arm}) for comparison are depicted in Figures 7.5 a) and b).

To generate these spectra for individual antennas, first a raster-scan was performed as seen above in Figure 7.3 a). Then, the excitation power was adjusted such that the specific antenna that was to be investigated could be clearly seen against the background, carefully avoiding melting or destruction of the structure. Finally, the scanning stage was positioned such that the laser directly excited the antenna, and the emitted light was then sent to the spectrometer instead of the APD. As the spectra were thus all obtained with different excitation powers, they were normalized to their peak emission. For all spectra, the integration time was 5 s, the excitation laser was polarized along

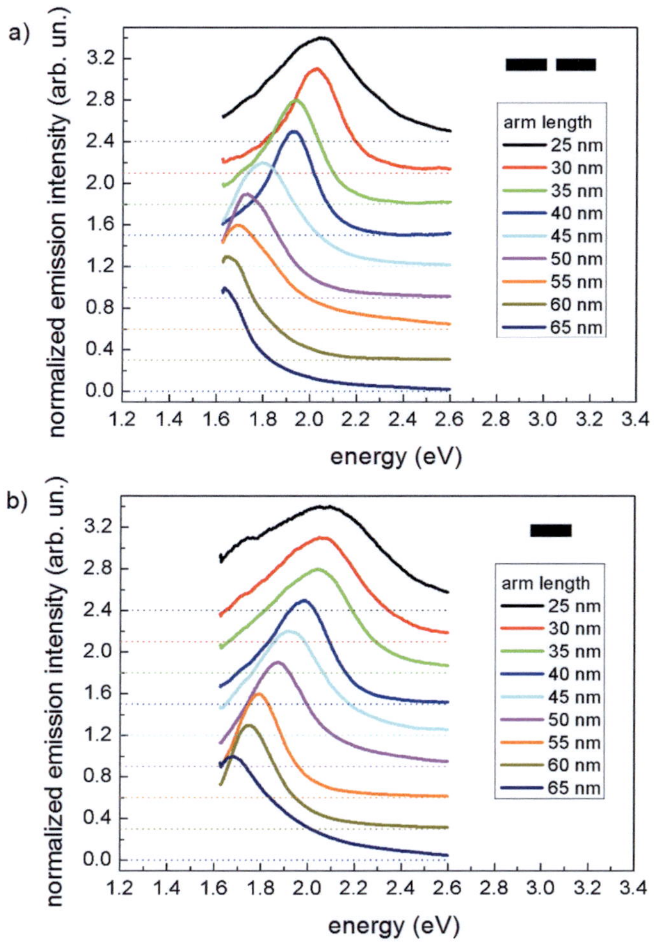

Figure 7.5: Normalized TPL spectra observed for a) gold two-arm antennas and b) the corresponding single-arm antennas for the polarization filter in the detection channel tuned to the longitudinal (long axis) polarization. Individual spectra are offset by a value of 0.3 to facilitate discrimination between them. Excitation was along the longitudinal axis only.

the long antenna axis, and the polarization filter in the detection path was set to let pass the longitudinal components of the antenna luminescence only. Two main results are obvious from the spectra. First, there is a red-shift of the peak emission for an increase in arm length both for two-arm and single-arm antennas, and second, a significant red-shift for two-arm structures compared to the respective single-arm counterparts. Both results directly resemble what had been observed for linear scattering experiments in the previous chapter. As already stated in the theoretical section, this red-shift for two-arm structures can e.g. be described by the hybridization model available for simple coupled dimers [136], where the luminescence peak intensities are attributed to symmetric, bonding modes of lower energy compared to anti-bonding dark modes.

Looking at Figure 7.5, one may notice that the spectra seem to broaden for the smallest structures under investigation, most pronouncedly for the single-arm structures. This broadening is due to substrate luminescence generated by the high optical intensities required for generating an optical response, especially for single-arm structures, which are less efficient in taking in the light. It is also evident that for longer antennas than shown here the spectral peak moves further out of the detection window limited by the frequency filtering required to suppress the excitation laser light in the detection channel. Even for longer nanoantennas (investigations were carried out up to 110 nm arm length), there was no spectral evidence for a resonance peak from a higher order plasmon mode (which might be optically allowed due to the nonlinear excitation scheme) such as an optical dark mode predicted by the plasmon hybridization model within the detection window.

It is now time to compare in greater detail the scattering spectra obtained in the previous chapter with the spectra generated using two-photon luminescence depicted in the last section. For an easy comparison, the scattering spectra from Chapter 6.1.1 have also been normalized and are presented in a similar matter as the ones from this section in Figure 7.6.

It can be easily seen that the spectra are very similar both in peak wavelength and resonance width. The comparison for the smallest structures is difficult due to signal-to-noise ratio problems for the scattering spectra and the superimposed glass luminescence observed for the TPL spectra. Nonetheless, the peak scattering energy comparison yields very convincing results, as plotted in Figure 7.7. Presented are the resonance peak energies for different arm lengths both for two-arm and single-arm nanoantennas as obtained from the two different excitation schemes. The manufacturing uncertainty of about ± 1 nm (which was transferred to a resonance energy uncertainty using the

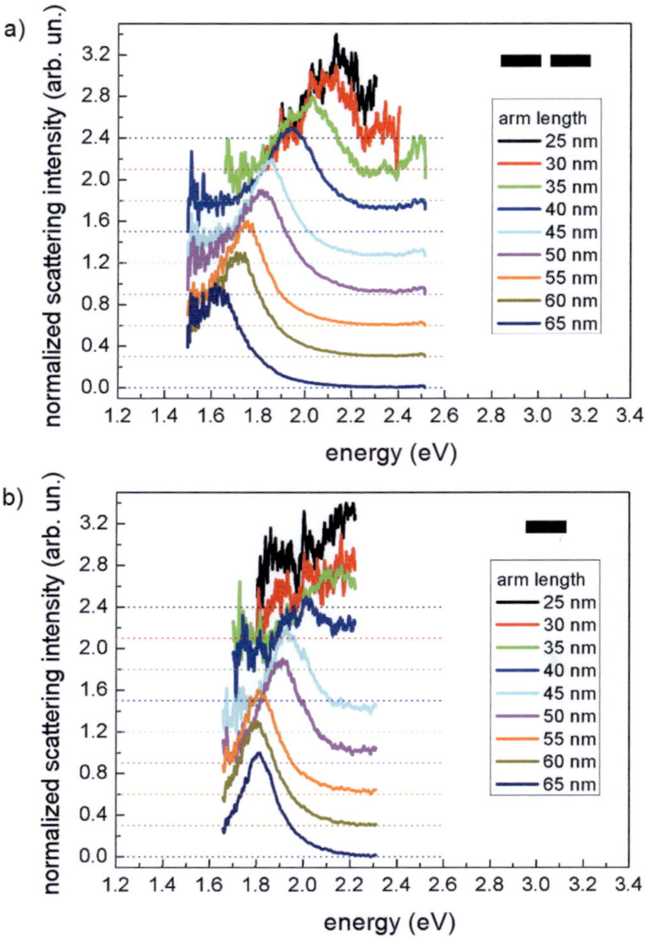

Figure 7.6: Normalized gold nanoantenna scattering spectra obtained by dark-field microscopy (same data as in Figure 6.5) for comparison to the TPL spectra shown above. Again the individual spectra are offset by a value of 0.3. Note the comparably low signal-to-noise ratio, and the similarity in curve shape and peak position to the TPL data.

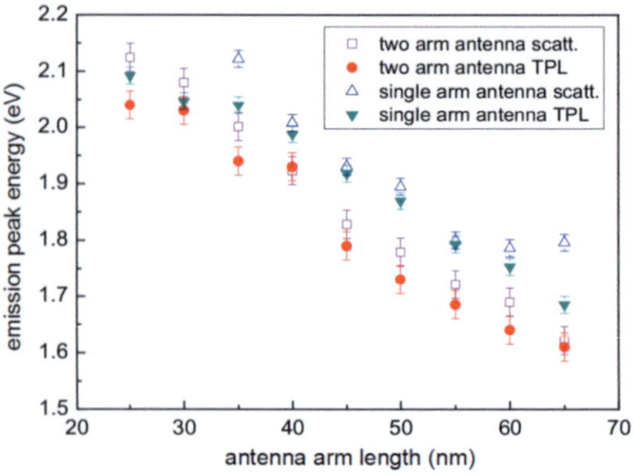

Figure 7.7: A comparison of the emission peak energies for antennas with the same nominal fabrication parameters under two-photon luminescence (TPL) and as observed using linear scattering. More details can be found in the text.

slope of the fitting curve seen in the previous chapter), and, for the two-arm antennas, an additional uncertainty derived from the strong coupling influence of the gap size, was used to generate error bars for the position of the peak resonance energy. The peak positions clearly show a very strong resemblance.

From this evident similarity of resonance peak position and spectral shape for single-arm and two-arm nanoantennas, it can be deemed very probable that the plasmon relaxation model by Dulkeith et al. [51] already presented in Chapter 2.4 not only holds true for single nanoparticles, but is still a valid description for processes in coupled structures, even under two-photon excitation conditions: A two-photon excited d-band hole generates a plasmon that subsequently decays radiatively. Since only one wavelength (or to be excact, a very small wavelength region around 810 nm) is offered to the antenna in this experiment, the observations made here also demonstrate that the typical scattering spectra of optical antennas, be they coupled structures or not, are indeed plasmonic eigenmodes, i.e. the wavelengths of the highest photonic mode density available. On a side note, such an observation is consistent with experiments on the specific emission enhancement of colloidal quantum dots (QDs) near silver nanoprisms with scattering spectra overlapping the QD emission spectra [129] or the use of silver nanoparticles to increase surface-enhanced Raman excitation spectroscopy efficiency [120] for molecular nanoanalytics.

7.3 Gold Antenna Transverse Plasmon Response

In the previous sections, the longitudinal response of gold nanoantennas has been investigated. As has been shown before, nanoantennas can exhibit not only longitudinal, but also transverse modes. In the chapter on scattering experiments, it was described that such transverse modes are very challenging to observe, especially if emitted from very small structures. However, with the two-photon luminescence excitation scheme, there could be a new way of detecting the transverse mode: Since the optical antenna two-photon induced luminescence is mediated by electronic states, the light emission does not require to be polarization conserving. In fact, once the energy has been transferred to electronic states, those states do not remember from which polarization state of the incoming light that energy has been deposited. Thus the emitted light should also contain photons which are transversely polarized, even with longitudinal incoming polarization (which has been shown to be the only efficient way for excitation). Again the emitted light intensity should correspond to a density of states, this time of the transverse modes, which indicates that the response should be weak in comparison to the longitudinal modes. Once the transverse mode has been found, its observation is a powerful tool to map out the gold antenna resonant arm length L for a fixed laser excitation wavelength λ using the cross-polarized transverse mode emission intensity, as the emission spectra should all lie completely within the spectrally available detection window.

7.3.1 Results and Discussion

The measurement setup is exactly the same as already depicted in Figure 7.1, except that the polarization filter in the detection channel, item (10), is now set to only let through the transverse antenna response. All antennas are still excited longitudinally, however, to allow for a large amount of energy deposition in electronic states. This time, the total integration time per spectrum was 30 s. As before, the laser excitation power was adjusted for each antenna length to yield a response with a favorable SNR avoiding the destruction of the nanostructures. Transverse luminescence spectra obtained for four two-arm and single-arm antennas each can be found in Figure 7.8 (smaller antennas are not shown as the amount of light emitted was too weak for the available destruction-avoiding excitation powers). As in the previous section, the response spectra have been normalized to unity to allow for an easy comparison of the spectral characteristics. Before taking a closer look at those, it is useful

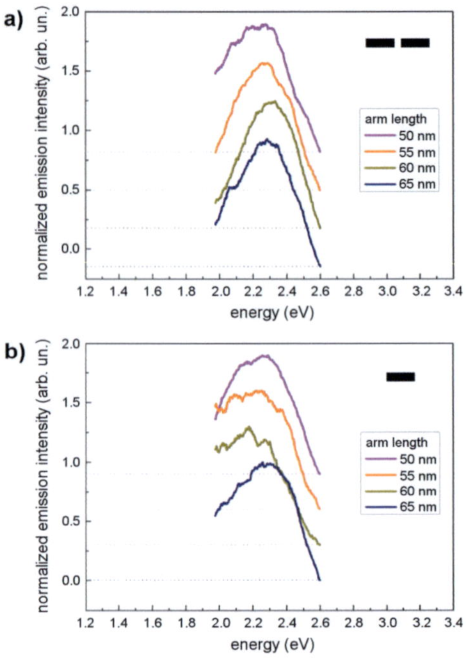

Figure 7.8: Normalized TPL spectra for the transverse polarization component of the light emitted for a) two-arm antennas and b) single-arm antennas. Excitation was along the longitudinal axis only. Once more the individual spectra are offset by a value of 0.3.

to compare the emission intensities involved: The transverse signal is about two orders of magnitude less intense compared to the longitudinal signal. But similar to the previous section, if a two-arm antenna were measured with the same excitation power as a single-arm structure, the former would be about 10 to 100 times brighter.

The transverse polarization TPL spectra for two-arm (Figure 7.8 a)) and single-arm (Figure 7.8 b)) antennas all show a resonance energy peak around 2.3 eV. Most notably, not only the peak energy, but also the spectral shape appears to be fairly independent of the long axis length (although a small aspect-ratio dependency cannot be ruled out) as well as gap coupling effects. The peak position is significantly blue-shifted to the longitudinal scattering resonances of the antennas that have been described earlier. This clearly suggests that the transverse plasmon mode of the structures under investigation is observed, as the antennas all have a shorter width than length.

As measured transverse scattering spectra are unavailable for a direct comparison due to the very low SNR, the best way to compare the peaks to what

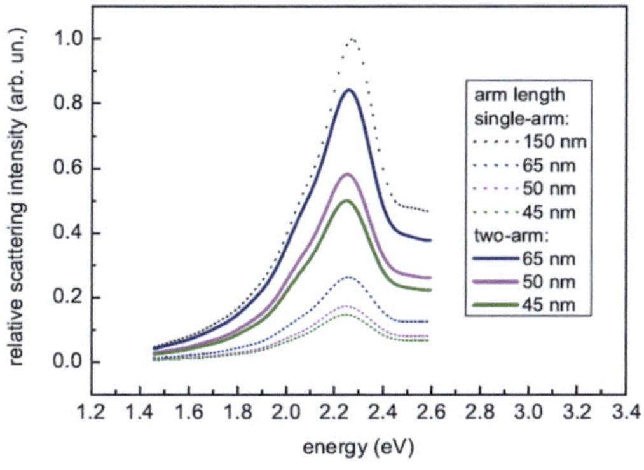

Figure 7.9: Numerical simulations of scattering spectra for single-arm and two-arm antennas with different arm lengths. The incident light was polarized along the short axis to obtain the transverse response.

would be expected for such experiments are numerical simulations. The following data and description on the simulation was kindly provided by Dipl.-Ing. Carola Moosman who works on the numerical modeling of optical antennas at the LTI. The FDTD-based software Lumerical [115] was used in this case. In these simulations, a plane wave with polarization perpendicular to the antenna length and parallel to the substrate surface was incident on gold nanoantennas of various sizes. The incident pulse encompassed the visible regime with wavelengths from 400 nm to 800 nm. Simulations were terminated when the energy had decayed to 10^{-5} times its peak value. To accommodate the experimental situation, the antenna arm width was chosen to be constant at 20 nm and the height fixed at 30 nm. The antenna was placed on a glass substrate (n=1.5) covered with a 30 nm ITO layer (which was described by a complex, frequency dependent dielectric function [125]). The dielectric function of gold was taken from the measurements performed by Johnson and Christy [89]. All material parameters were fitted to a generalized multi-coefficient model supported in the FDTD simulation. To avoid extremely high field enhancements at singularities, the corners and edges of the antenna arms were rounded with a radius of 3 nm.

Within a simulation area of 800^3 nm^3 surrounded by 12 perfectly matched layers (PMLs), the mesh cell size varied between 0.15 nm and 8 nm throughout the simulation area. The plane wave was incident onto the antennas from inside the glass substrate. A power monitor situated in the glass substrate 100 nm above the antenna detected the scattered power. The latter was then normalized by the incident intensity and is thus proportional to the power scattered backwards into the glass substrate. Figure 7.9 shows the simulaton results. Relative scattering intensities of single-arm and two-arm antennas with arm lengths of 45 nm, 50 nm and 65 nm and a further single-arm antenna with an arm length 150 nm are depicted. Evidently, the peak position at 2.25 eV is in very good agreement with the measured results presented in Figure 7.8. As for the measured results, the peak resonance energy shift is practically unnoticable. With an increase in antenna volume, the scattering response simply increases accordingly. This makes the observation of the transverse mode even more likely.

As explained in the introduction to this section, the energy transfer model by Dulkeith et al. [51] would specifically allow for the non-polarization-conserving luminescence that has been observed. Again it can thus be confirmed that the model holds true even for structures under two-photon excitation. It is also to be noted once more that the TPL experiment allows for a better SNR than scattering experiments, which makes it a very useful characterization technique aside from the physical implications of the eigenmode relaxation observation.

Another interpretation of the observed cross-polarized signal would be that it is material-intrinsic and not due to the antenna resonance, as similar spectral responses have been observed for chemically synthesized gold nanorods under two-photon excitation by e.g. Imura et al. [85] and Scolari et al. [167]. For the structures presented here, which are polycrystalline, such a purely material-intrinsic response is unlikely, however. The non-resonant TPL of gold nanostructures which were fabricated similarly, which will be presented in Chapter 9 and which doesn't include any such L symmetry point luminescence contribution (which would appear in a similar spectral range around 2.3 eV), makes it even less probable that this interpretation is valid. A means to shed more light on the interpretation would be to conduct comparative electron energy loss spectroscopy (EELS) measurements both on single crystalline and polycrystalline structures [22, 134, 67].

In the analysis of the longitudinal emission, it was stated that the perfectly resonant antenna could not be judged on its intensity, as large parts of the emission were in the same spectral range as the exciting laser that needed to be filtered. Now, with the selective observation of the transverse plasmon mode, a complete mapping of the resonance features is possible. The transverse intensity, which lies completely within the detection window for all structures, can be used as a direct (but still nonlinear) measure for the strength of the antenna resonance. As performed previously, an integration of raster scan pixels

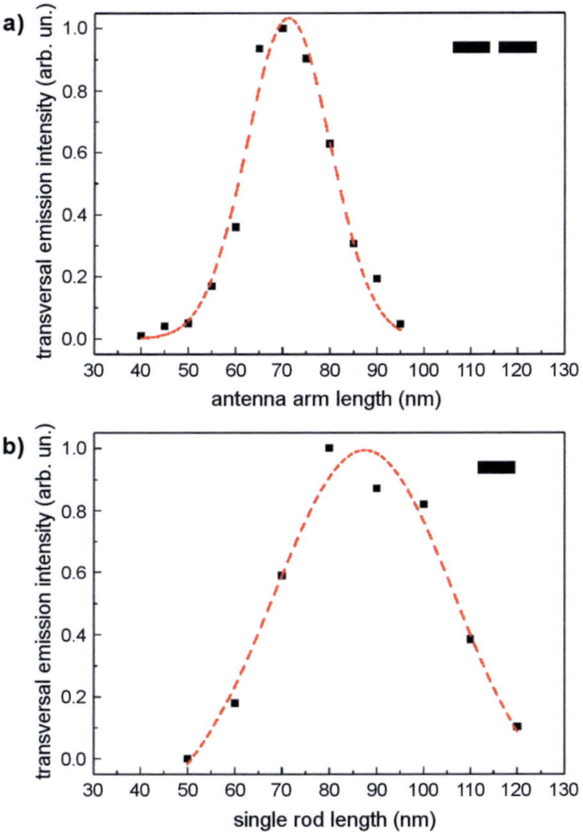

Figure 7.10: The normalized integrated luminescence photon counts from the transverse plasmon response for different antenna arm lengths under investigation. a) gives the result for two-arm antennas, b) depicts the intensity from single-arm antennas. The dashed line Gaussian fit is a guide to the eye.

from various antennas gives total photon counts emitted by specific antennas for a constant, non-destructive excitation intensity. Such total counts, normalized to the maximum emission photon counts from the most resonant antenna, are mapped out for two-arm and single-arm antennas in Figure 7.10. Evidently, it is still the absorption efficiency mediated by the excitation and thus the longitudinal arm length resonance that is mapped out (again similar to the experiments conducted by Wang et al. [182] already cited earlier), while the energy is transferred to the transverse mode for the luminescence response according to the available photonic density of modes. While the position of the maximum for both curves is independent of the excitation intensity, the curve shape is not, as the excitation process is nonlinear and thus e.g. a lower excitation intensity would lead to a flatter curve. Nonetheless, a Gaussian fit is shown alongside the actual measurement data points as a guide to the eye. Even though the single-arm antenna curve was generated for a slightly higher excitation power, the curve is still flatter. Assuming a comparable focus quality, it can again be concluded that the excitation efficiency for those structures is much lower.

The resonant structures for two-arm antennas are found to have an arm length of about 72 nm, while the corresponding single-arm antennas show a resonance maximum for a length of about 87 nm. Both those values are in very good agreement with the extrapolation from the known peak wavelengths for shorter structures obtained from linear scattering experiments already shown in Figure 7.2.

Cross-polarized TPL measurements provide a powerful tool to quickly characterize antenna resonance properties. As many other excitation wavelengths could be used, it is perfectly suited to find a resonant antenna for a specific desired wavelength. On a side note, it is well-known that high numerical aperture objective lenses under linear polarization cause the excitation light to become slightly depolarized, see for example reference [139]. This cannot be the cause for the transverse luminescence observed here, however. Not only didn't the antenna show any response for excitation light that was mainly polarized along the short antenna axis, but if that were the case, the emission signal observed should become more intense for larger structures as well, which is clearly not the case.

7.4 Aluminum Antenna Plasmon Response

In addition to gold nanoantennas, aluminum structures were investigated. Two-arm as well as single-arm antennas of the same size and also larger ones (as it is already known from the previous chapter that antenna resonances are heavily blue-shifted for the same antenna size and thus resonances in the visible occur for longer antennas for aluminum compared to gold) were closely examined. Just as in the chapter on scattering, the antenna width was increased to 30 nm to allow for an acceptable signal-to-noise ratio.

The excitation and detection conditions were the same as described above for gold nanoantennas. The detector for the spectral response was changed to a Jobin-Yvon Horiba nitrogen-cooled CCD camera (integration time of 8 s). For intensity investigations, still the single-photon detectors were used.

Again, at first the excitation properties for 810 nm pulsed laser light were examined. As for the gold antennas, only excitation with light polarized along the longitudinal axis led to a significant antenna luminescence as detectable by the single photon detectors. The transverse excitation again showed no noticeable response. If the same excitation mechanism as for gold is assumed, this is easy to explain, as the transverse mode resonance for aluminum antennas of 30 nm width is even more severely detuned from the excitation wavelength than it had been for gold antennas. The excitation laser polarization was thus fixed to be along the longitudinal axis for all experiments described in the following.

7.4.1 Emission Intensity Investigations

As for gold nanoantennas, a clear increase in antenna response could be observed for an increase in antenna arm length from very short antennas towards the ones known to be resonant closer to 810 nm from Chapter 6. Of course, 810 nm is possibly not the optimal wavelength to excite such structures, as the wavelength is located in the interband transition regime of aluminum as also observed in the scattering experiments, where a peak wavelength of 810 nm (1.53 eV) could not be reached. Nonetheless, in the following it will be seen that the experiments can still be carried out successfully. For gold antennas, it has already been discussed that it is best to investigate the intensity emission of the transverse mode in the detection channel to find resonant structures. There is, however, only a very weak transverse emission for aluminum nanoantennas (too weak to be investigated spectrally), even for antennas strongly emitting in the longitudinal mode. There are two main possible explanations for this behaviour: Possibly the transverse intensity is either largely emitted in the UV (due to the small width of the antennas), where the single-photon detector used has a very low detection efficieny, or the density of states available in the transverse mode for structures that thin is too small compared to the one for

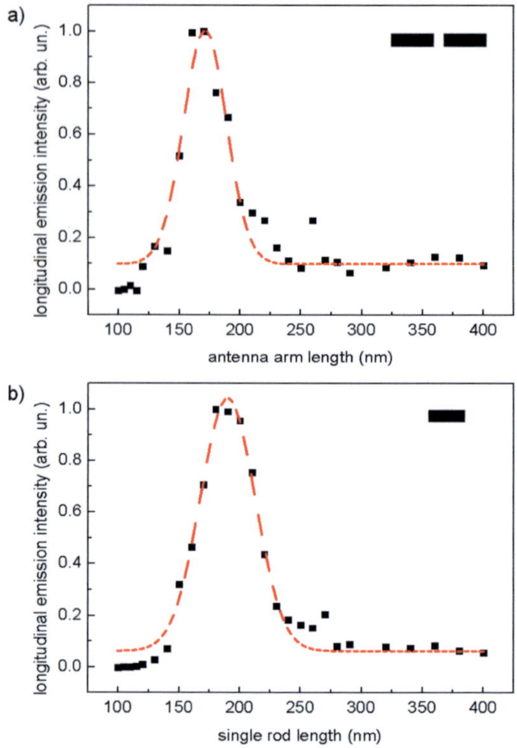

Figure 7.11: Integrated normalized multi-photon emission from two-arm and single-arm aluminum antennas of various arm lengths. The dashed red line Gaussian fit is a guide to the eye. The investigation of single-arm structures was carried out at about four times the power used for two-arm structures.

the longitudinal mode. An additional study with wider antennas (a width of 50 nm was checked and still proved insufficient) could give additional insight into that behaviour.

For this study, however, thus the longitudinal mode intensity is available exclusively. As for gold antennas, the emission intensity is only faithfully reproduced up to the filtering wavelength of the excitation laser if one assumes again that the emission is spectrally shaped by particle plasmon properties. On the other hand, it is expected from the spectral investigations and theoretical predictions considering the dielectric function of aluminum that the scattering peak starts to remain in a relatively constant range around 670 nm (1.85 eV) both for two-arm and single-arm structures. Thus it may still be interesting to take a look at the entire *arm length resonance curve* for a fixed excitation wavelength, which is given in Figure 7.11 for two-arm and single-arm structures. Analogous to the scattering spectra, where there was a red-shift for the spectral peaks of the two-arm compared to the single-arm structures, the single-arm peak in the curve is found to be at a larger arm length (190 nm) compared to the two-arm structures (170 nm). However, this observation may still be strongly influenced by the filter in the detection path. The resonant arm-lengths are thus only considerable approximations, which will become even more obvious when the spectra are observed.

It is of course also of interest to determine the type of interaction observed for the aluminum luminescence. Is it again a two-photon process that is observed? To obtain this information, an attempt was made to observe a power curve. These attempts were not fruitful, however, as aluminum structures tend to be destroyed much more quickly and before emitting great amounts of light under the laser illumination. This is somewhat understandable as they are driven at a wavelength which directly feeds interband transitions which should favor non-radiative decay and in turn may lead to additional heating. It is also possible that in general luminescence processes in aluminum, which have not been studied for single aluminum antenna structures so far, involve a greater amount of non-radiative energy deposition compared to light emission. The multi-photon processes involved thus require an additional detailed study.

7.4.2 Spectral Investigations

Despite the non-advantageous melting properties, spectra generated from laser excitation could be observed. As for gold nanoantennas, these aluminum luminescence spectra have been normalized (due to different excitation powers for different types of structures) and are depicted in Figure 7.12. Differing from the scattering spectra, there is no visible effect of the interband transition on the emitted mode down to the filter energy of 1.6 eV. From smaller to larger structures, the red-shift simply continues. It is thus more difficult for aluminum antennas to directly identify the emission with the spectra as observed via scattering in the previous chapter. In contrast to the peak positions, the spectral width is very similar to the scattering results: It is again about 0.5 eV for all structures observed. Possibly because non-radiative channels are more dominant for aluminum structures, the spectra can only be depicted for comparably large structures.

A trend on the peak wavelengths that could be observed can be given and compared to the scattering results from the previous chapter (again results from structures from different samples, but with the same fabrication parameters are

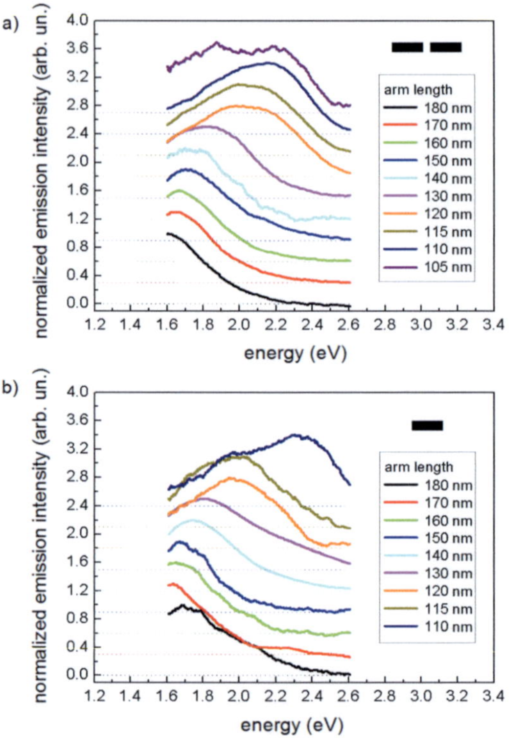

Figure 7.12: Normalized aluminum multi-photon luminescence spectra observed for a) two-arm antennas and b) the corresponding single-arm antennas for the polarization filter in the detection channel tuned to the longitudinal (long axis) polarization. Excitation was along the longitudinal axis only.

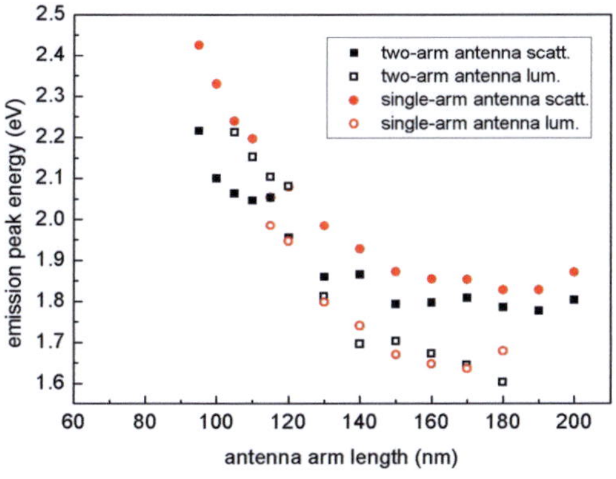

Figure 7.13: Comparison of the peak wavelengths obtained from multi-photon excitation and from scattering experiments for two-arm and single-arm aluminum nanoantennas. The peaks for arm lengths yielding scattering results within the interband transition clearly differ for scattering and multi-photon generated luminescence.

compared). This is depicted in Figure 7.13. It is evident that the emission cannot be matched as easily to the scattering peaks as had been the case for the gold antennas. Possibly, melting effects already induce structural changes even though spectra can still be recorded, especially for smaller structures. The values below the interband transition energy may possibly be explained by assuming another emission scheme, where due to the nonlinear excitation scheme new pathways for relaxations are opened.

7.5 Conclusions

It was demonstrated with a detailed study that for gold nanoantennas both longitudinal and transverse two-arm and single-arm antenna response spectra obtained by two-photon illumination follow the spectral shape of the photonic eigenmodes, i.e. the particle plasmon. This is particularly interesting as the white-light continuum spectra observed by Mühlschlegel et al. [128] or very recently by Huang et al. [83] indicated a different behavior. For bow-tie optical antennas, broadband continuum emission was reported as well by Schuck et al. [162]. As these studies didn't use structures much different from the ones investigated here and there also is a strong similarity in terms of the laser systems and optical setup used, the typical powers applied ($\approx$50 μW [128, 83],

≈5 μW [162], this work: ≈10 μW) and the laser pulse duration (≈1 ps [128, 83], ≈120 fs [162], this work: ≈500 fs), a different explanation for the discrepancies needs to be found. Some differences may be explained by the use of focused ion beam milling in references [128, 83], by which gallium might be deposited in or near the antenna gap. This can not be the only explanation, however, as Schuck et al. did not use that technique.

For aluminum antennas, multi-photon excitation can be performed as well. However, structures are much more easily destroyed (possibly due to the excitation at a wavelength within the interband transition region), which makes spectral measurements more challenging. While the spectral width observed in scattering experiments is reproduced using nonlinear excitation, the spectral peaks are more difficult to compare. For longer structures with constant scattering peak energies around the interband region, differences occur between luminescence and scattering spectra. Further investigations, for example using electron energy loss spectroscopy, might yield additional insight into this phenomenon.

For all structures, whether made of gold or aluminum, the scheme can be interpreted as a largely tunable nonlinear frequency conversion process. Especially with emission from the transverse mode for gold, antennas with very efficient excitation cross-sections as well as significant emission from the plasmonic mode can be designed, thus not only generating frequency doubling or fourth-harmonic generation, but an inherently freely choosable wavelength transformation. Of course, one needs to operate within the suitable wavelengths outside of absorption maxima for the material chosen.

This technique can also find very practical applications where a resonant antenna needs to be found for a specific excitation frequency. Most notably, it can easily be combined with setups for atomic force microscopy or tip-enhanced near-field measurements. Such studies could be used specifically to find where the near-field is most intense around the optical antenna structure. Simulations imply that this near-field enhancement should be greatest within and near the antenna gap, an assumption that is corroborated by the fact that for gold a great TPL emission enhancement could be observed for the two-arm antennas with gap compared to single-arm structures of the same size. First steps into this direction will be shown in the next chapter, where near-field effects of gold optical antennas under laser excitation will be investigated.

Chapter 8

Near-Field Measurements

In the previous sections, two far-field characterization methods for optical antennas were outlined, namely dark-field microscopy and two-photon luminescence measurements. A different approach to learn more about the properties of such nanostructures is near-field characterization. In the following, an approach will be presented to measure near-field effects in optical gold antennas using an apertureless scanning near-field optical microscopy (a-SNOM) method. As the most notable result, the TPL observed in the last chapter is shown to be significantly correlated to gap properties.

Apertureless SNOM is a near-field characterization method relying on the scattering properties of a small probe of subwavelength extent, for example the tip of an AFM cantilever, which approaches a sample which is raster-scanned. The tip thus scatters out light that would not be able to propagate to the far-field (and is thus invisible to far-field investigation approaches) if the tip were not present. The tip is commonly placed in a fixed position on the optical axis in the focal plane of the objective lens and illuminated by laser light, while the sample is raster scanned below. In general, however, if the probe is raster-scanned over the sample, the presence of such confined light at different points can also be investigated (especially for structures of subwavelength dimensions), as shall be shown below. Figure 8.1 exemplifies this approach.

A-SNOM allows to understand more completely the nature of electromagnetic responses of nanostructures. Beside nanoaperture based approaches [174, 152, 110] and scanning of single fluorescent molecules or quantum dots attached to a probe over the sample [59, 100, 175], the usage of a scattering [96, 25, 181] or field enhancing [79, 172] tip has become a standard technique for probing highly localized electromagnetic fields. Typically, the electric field component is accessed, but even the magnetic field component can be probed [32]. Often, additional demodulation of the information obtained is

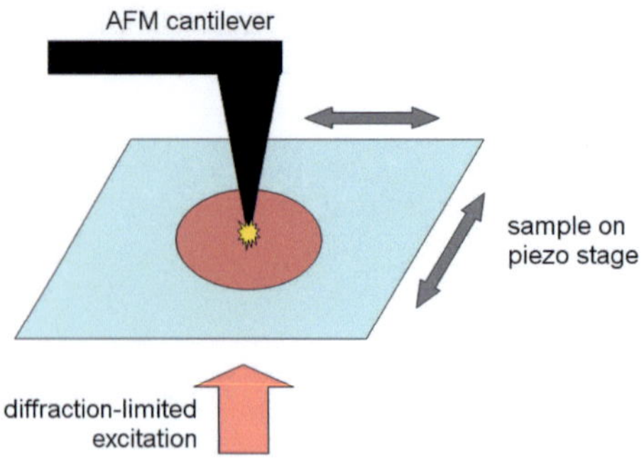

Figure 8.1: Working principle of typical a-SNOM setups. A scattering probe (black) with a very small tip, which is often metallic, but can also be made of a dielectric, is placed in the center of a diffraction-limited laser focus (light red). Then, the sample is raster-scanned in the focal plane. If enhanced near-fields are present, they interact with the tip (yellow) and can partly propagate to the far-field. If the response is then collected for each point, a sub-diffraction resolution image is obtained. Often, the response is even more increased by additional enhanced fields forming at the tip.

performed to separate the comparably faint near-field response from the far-field background [117]. Alternatively, spectral filtering to detect the second-harmonic enhancement component only is applied [24]. Of particular interest are near-field images of nanometer-sized metal structures such as optical antennas [128, 59, 137, 133, 93, 176]. These images can give additional insights as to which factors determine the antenna plasmonic resonance, which is crucial for applications such as biosensing or efficient molecule coupling. Large fluorescence enhancement [93] or Raman cross sections [197] have been observed. While two-photon induced luminescence (TPL) based far-field microscopy is an interesting alternative for such characterization [71], the best resolution for such direct optical mapping is provided by tip-based approaches.

In this chapter, an inherently demodulation-free a-SNOM approach where the tip is raster-scanned over the sample and the objective lens is presented, where the intensity of the optical response of the nanostructure, which is kept in a fixed position within the laser excitation volume, is observed. Using this comparably simple approach, near-field measurements of the optical two-arm antennas with a 20 nm gap already investigated in the previous sections, will be generated. Compared to previous SNOM approaches which worked with infrared (IR) or mid-infrared wavelengths [161, 143], the antennas observed here are again driven using a pulsed laser operating at 810 nm and thus operate very close to the visible spectrum. Also, in contrast to pseudo-heterodyne approaches [161, 143], the intensity information from the tip-antenna inter-

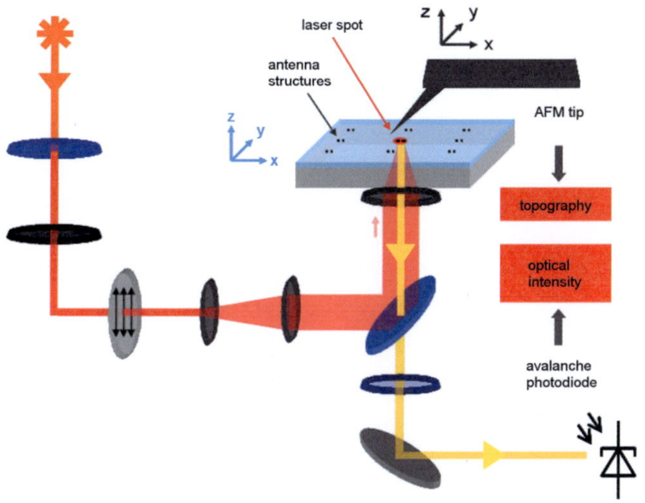

Figure 8.2: Experimental scheme for the near-field characterization. Detailed explanations are given in the text.

action (photon counts) is directly observed. In comparison to measurements where the near-field intensity of optical antennas of similar dimensions on top of a laser diode at near-IR wavelengths was observed [40, 195], this method works through the excitation with focused laser light and is thus applicable for the characterization of free-standing, non-integrated structures. Additionally, differing to all three approaches, it relies on the observation of light generated by the nanostructure, mediated by a two-photon excitation process, which allows for simple spectral filtering of the excitation light.

8.1 Experimental

The measurement technique used is basically the same that has already been applied to generate the two-photon luminescence response in the previous chapter. In addition, however, the effect of an AFM cantilever approaching and interacting with the nanostructure, is observed. Thus the light emitted solely by the structure and/or through the interaction with the AFM tip (Nanosensors ATEC-FM (doped silicon), originally designed for force modulation mode, used here in intermittent contact mode at oscillation frequencies around 70 kHz) placed on top, is investigated. The setup is shown in Figure 8.2. Both sample stage and AFM cantilever can be moved independently in all three spatial dimensions via piezo crystals, which enables raster-scanning (JPK Instruments TAO-stage). The photon counts are first converted to TTL pulses at the detector output and finally to a DC-voltage by integration via the internal capacitance and inductance of the AFM controller auxiliary input channel. The AFM can be used either in intermittent contact or in contact mode. In intermittent contact mode, the cantilever oscillates sinusoidally near its resonance frequency (for the cantilevers used here with an amplitude of about 60 nm). Changes in surface height are detected by changes in oscillation frequency. For AFM imaging, intermittent contact mode has the advantage that the cantilever is rarely very close to the surface, which makes it less likely that cantilever or sample are damaged. In contact mode, the cantilever operates directly on the surface, and surface height is detected by bending of the cantilever. In this case, the average force exerted on the sample is generally larger.

The a-SNOM measurement is conducted as follows. After initial setup and tuning of the cantilever, the latter is aligned such that it can be easily positioned close to or over the center of the objective lens, which acts as a point of reference for both sample stage and the AFM piezo crystals. Then, a sample with several structures is raster scanned over the laser excitation spot, with the laser properly linearly polarized for the maximal response, i.e. along the long antenna axis. As seen in the previous chapter, an optical response is generated by each nanostructure due to this excitation (see also e.g. references [128, 70]), which is essentially diffraction limited. This response is most significant for resonant structures, as the coupling of the field to the structure is most efficient in that case. An enlarged spot observed in the far-field for a typical antenna structure is shown in Figure 8.3. The full width at half-maximum (FWHM) was about 350 nm, as expected for a two-photon process (the emission intensity depends quadratically on the laser excitation power [189]) observed with a high NA objective lens.

For a near-field measurement, a single structure is then chosen and placed over the objective lens in a fixed position, such that a (far-field) optical response with constant intensity over time is observed. To avoid laser-induced destruction of the structure and saturation of the APD, the laser intensity is adjusted at this point to typically about 0.5 mW in front of the objective lens entrance pupil. Then, the AFM tip is placed on the sample near that structure and, in intermittent contact mode operating at an oscillation frequency of about 70 kHz, it is raster scanned over the sample. An AFM image yielding topography information is thus generated. At the same time, while the tip is moving, the APD detects the currently measured optical response, which changes with the position of the AFM tip and thus measures the superposition

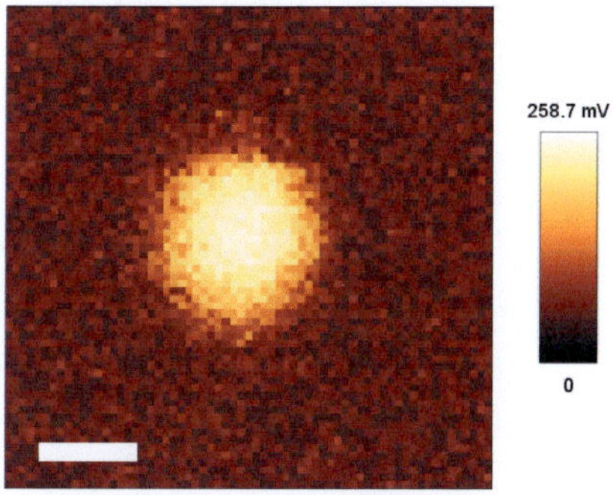

Figure 8.3: View of a typical diffraction-limited spot for an 85 nm arm length two-arm antenna. The units of mV are proportional to photon count numbers, the scale is linear. Scale bar length is 200 nm.

of the far-field background and the response that is additionally generated by the near-field interaction. As the tip is tilted by 45 degrees, a superposition of in-plane and out-of-plane near-field components is observed. To keep the effects as symmetric as possible, the tip is aligned such that its projection on the sample is perpendicular to the long antenna axis. It was checked that the laser intensity is low enough that no optical response can be observed emanating from the cantilever itself when it is scanned over the objective lens without a nanostructure present. No demodulation with the tapping mode frequency is performed, the data is shown as it was measured over the average interaction time, i.e. for sinusoidal oscillations of tip-sample distance. The far-field background described previously is inherently present in all measurements depicted in the following. However, it is constant for a sample at a fixed position and thus directly subtracted from the near-field images presented in the next paragraphs. Additional details of the measurement procedure can be found in the *Diplomarbeit* of Gabor Varga [179].

8.2 Results and Discussion

A set of data combining scanning electron-beam microscopy SEM, AFM and near-field information for a two-arm gold antenna of arm length 105 nm (which is thus already significantly above antenna resonance for the excitation wavelength used, which in the last chapter had been shown to be about 85 nm) observed using the a-SNOM method is shown in Figure 8.4. In a), an SEM image of the structure recorded is given; b) and c) show the corresponding AFM image (recorded in intermittent contact mode) and near-field emission

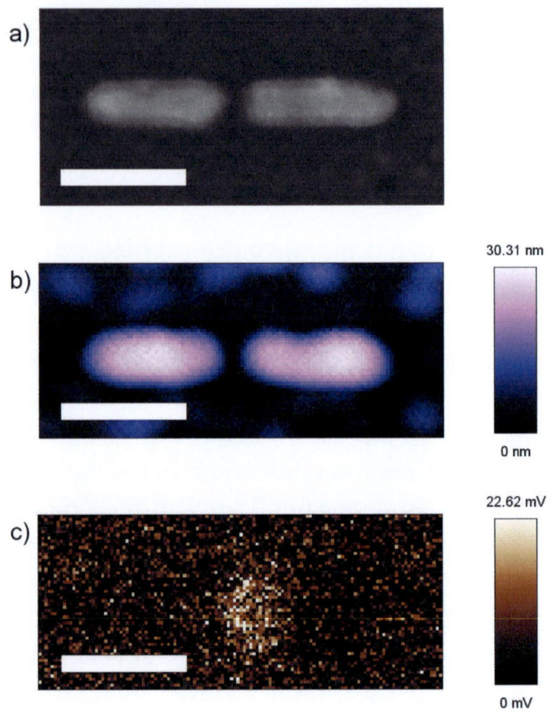

Figure 8.4: a) A scanning electron-beam microscopy (SEM) image of a 105 nm arm length antenna. b) Corresponding AFM image recorded in intermittent contact mode. c) Optical intensity image simultaneously recorded with b). Scale bar length is 100 nm in all images.

generated by the interaction, respectively. The near-field information is given in units of mV after the pulse-frequency DC-voltage conversion. A marked emission contrast for the presence of the tip near the gap of the structure is clearly visible.

A second set of data for a 120 nm arm length gold antenna (Figure 8.5) gives a comparison with the application of the previously presented scanning algorithm applying the intermittent contact mode (b,c) and the use of contact mode AFM raster scanning (d). The latter shows a substantially improved contrast, and gives (while the response in the middle of the gap already saturates the detector response) an additional enhanced response on the extremities of the structure. Please note that the cantilever used for this experiment had already experienced heavy use, which explains the rather poor quality of the AFM topography image. The significantly increased SNR for contact mode operation can be explained by the larger time the cantilever then spends very close to the surface. So it seems that this near-field method is preferably to be performed in contact mode, as long as the structures aren't damaged by the interaction with the cantilever tip, which is unfortunately more likely to happen than in contact mode due to the larger average forces exerted on the surface. The relatively large spot sizes near the gap in the near-field image are present as the center regions needed to be overexposed in order to make the enhancements at the antenna ends visible as well. The measured near-field pattern shown in d) constitutes the theoretically predicted pattern for the near-field components of a resonant optical antenna. A COMSOL Multiphysics® electromagnetic near-field simulation for the antenna geometry used is shown in e). The antenna nominal sizes of arm length 120 nm, arm width 20 nm, arm height 30 nm, and gap width 20 nm were used for this simulation, the antenna edges were slightly rounded (radius 3 nm) to avoid singularities. The inciding light had a wavelength of 810 nm and was polarized along the long antenna axis. The refractive index of gold was taken from the values of Johnson and Christy [89], and the 30 nm indium-tin-oxide layer was assumed to also have a complex refractive index [125]. As can be seen from both images, there is an increased near-field intensity near the outer ends of the rods, and an even more pronounced peak in optical intensity in the vicinity of the antenna gap. Notably, the experimentally observed enhancement in the gap is higher than near the ends of the rod in the measurement – however, the contrast is lower than would be expected from the simulations. This can be attributed to the limited ability of the tip to penetrate the gap as the size of the tip cone is on the order of the gap dimensions, as well as to the dynamic range currently available with the setup.

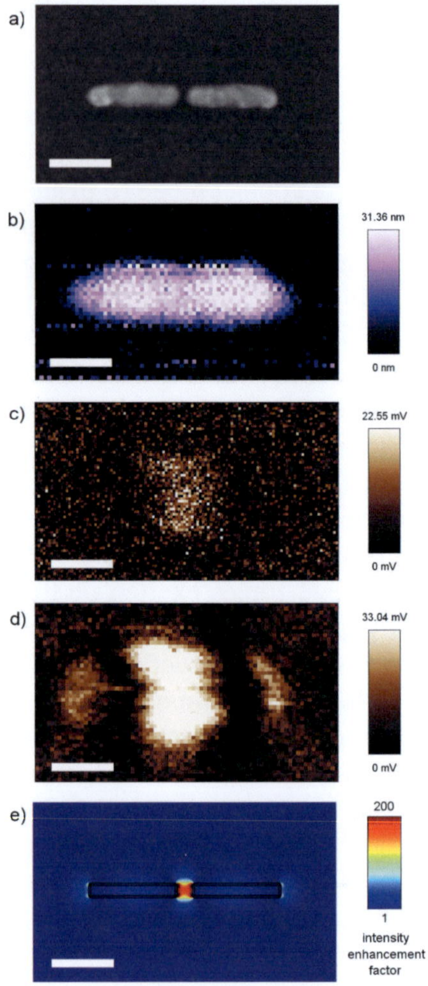

Figure 8.5: a) SEM image of a 120 nm arm length antenna. b) AFM image recorded in intermittent-contact mode. c) Optical intensity image simultaneously recorded with b). d) Optical intensity image recorded while scanning the cantilever in contact mode. e) COMSOL Multiphysics® simulation of the absolute value of the *in-plane* component of the scattered electrical field squared along the long antenna axis. Note that the intensity enhancement factor is several orders of magnitude higher for highly resonant structures. All intensity scales are linear. Scale bar length is 100 nm in all images.(Simulation results courtesy of Carola Moosmann)

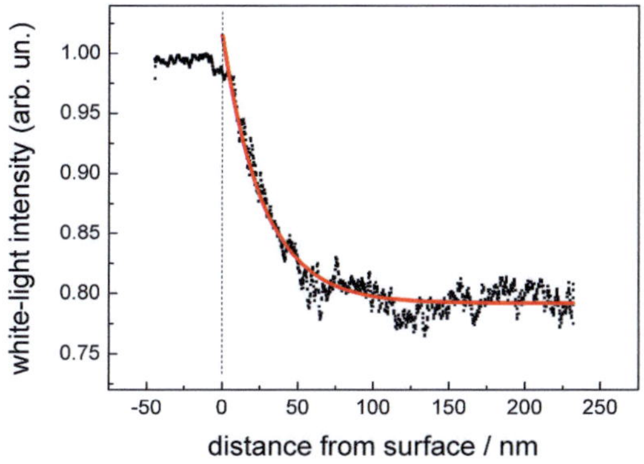

Figure 8.6: Optical intensity recorded for various distances of the cantilever tip from the antenna gap (black dots) and and an exponential fit to the data

To prove that really a near-field interaction is measured, it is essential to examine the dependence of the additionally observed light on the distance of the cantilever tip to the sample. To this end, the cantilever was slowly approached onto the sample right over the gap of a two-arm nanoantenna. The observed intensity for different distances was measured. The approach curve obtained is shown in Figure 8.6. For large distances (greater 100 nm) from the sample, only a constant far-field intensity is observed. Then, around 50 nm from the surface, the photon intensity starts to increase, to finally reach a maximum when it is in contact with the structure. The position where half the maximum intensity is reached is around 27 nm from the sample surface, which thus clearly indicates a measured near-field effect, similar e.g. to the optical tip-sample gap calibration for floating tip nanolithography [123].

8.3 Conclusions and Outlook

It could be shown with two examples that near-field components of radiation emitted by gold nanostructures which generate an optical response can be mapped out and characterized using this a-SNOM method where the emission intensity response is observed while the sample position is fixed with respect to the optical field and the tip of an AFM cantilever is moved over the structure.

While demodulation with the intermittent contact mode frequency is not generally necessary for the method to be successful, the significantly enhanced signal to noise ratio obtained for a measurement in contact mode suggests that the effectiveness of our a-SNOM setup can be increased even more by application of a lock-in amplifier for demodulation of the signal. This could also serve to improve the quantitativeness of the data presented. As a next step, it would be interesting to quantitatively compare the near-field information of antennas with different arm lengths at fixed excitation power to yield an even more thorough insight into the mechanism of optical generation, which might e.g. be possible with the aforementioned specifically fabricated AFM tips. Also, the change in luminescence generation by a single antenna for a detuning of the laser wavelength or a change in gap width could be investigated. A thorough spectral analysis of the light generated specifically at the gap (e.g. by spectral comparison of light generated with and without the tip) would be beneficial.

Seeing these successes, the next logical question to ask is whether there is a clear maximum of light concentration in the gap for further antennas, especially resonantly driven ones. There are a few difficulties involved in the execution of this method which have rendered it impractical to apply it for a great number of structures.

The greatest challenge is the non-reproducibility of tip properties. Each commercially available AFM tip, designed for surface measurements, has a slightly different tip shape. Unfortunately, this generates a very large difference in the quality and type of the optical response signals generated, and each tip again only lasts for a few measurements before it becomes unusable due to deformation. The best way to overcome this dilemma would be to fabricate AFM tips with antennas of their own on top of them with reproducible properties. Finally, adequate laser powers are required to generate the white-light response in the first place. As shown in Chapter 7, the power needs to be high enough for the generation of an observable signal, but low enough to avoid a deformation of the structure. This balancing act becomes even more difficult once the tip generates a near-field enhanced response – suddenly the powers may be too large. Possibly structures of greater width might provide higher stability.

Chapter 9

Lightning Rod Effect on Long Single-Arm Antennas

In the previous chapters, properties of resonant single-arm and two-arm antennas under highly focused laser illumination and their two-photon luminescence response have been investigated. Additionally, this two-photon luminescence was pinpointed to specific areas on the nanoantenna surface for such devices via apertureless scattering near-field microscopy. Now, laser induced far-field characterization results on non-resonant structures will be given, where lightning-rod effects (i.e. highly enhanced fields at structure singularities) and shape transformations are observed. Two distinct approaches are followed to gain insight into the behaviour of the rods: First, a two-photon luminescence process using a pulsed laser as an excitation source is investigated, with methods very similar to the ones used in Chapter 7. Specifically, the interaction cross section of the luminescence process and the emission spectrum will be observed, and investigations towards shape morphology changes will be carried out. In a second part, fields in the vicinity of the structures will also be investigated using a layer of quantum dots as local luminescent electromagnetic field probes. Parts of this chapter have previously been published in *Nanoscale* [190].

9.1 Two-Photon Luminescence Investigation

The gold nanorods used for the experiments were all produced using electron beam lithography exclusively, as were the nanostructures seen in the previous chapters. Again, all structures were spaced with comparably large distances to avoid any near-field coupling. An SEM image of the nanorods used of length

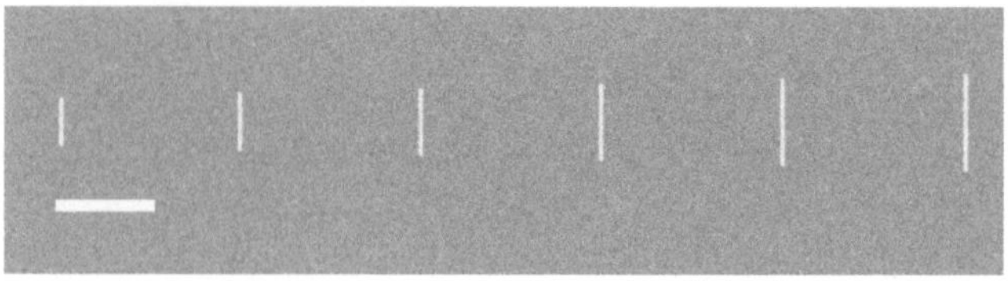

Figure 9.1: SEM image of several gold nanorods (length 500 nm through 1 μm in 100 nm steps) fabricated using the standard protocol as outlined in Chapter 4. Scale bar length is 1 μm.

1 μm as well as of shorter lengths (in 100 nm steps) down to 500 nm is shown in Figure 9.1.

As for the TPL of resonant structures, a strongly focused, femtosecond pulsed Ti:Sa-laser beam is used for excitation (this time tuned to 800 nm center peak wavelength). The luminescence response is again obtained after filtering of the excitation laser using the same short-pass filter as in Chapter 7. The laser is also raster-scanned over the nanostructures, this time larger nanorods of dimensions of several hundred nm x 50 nm x 30 nm (*L* x *W* x *H*). As shall be seen, a very strongly localized white-light response from the ends of the rods is obtained, which will be shown in the subsequent sections to be of high nonlinear order in terms of its spatial extent. The optical spectrum of the white-light that is generated from this non-resonant process is given, which beside a broad peak contains a second-harmonic frequency component.

The atomic force microscope already used for near-field characterization in Chapter 8 was used to check effects of the optical excitation on the morphology of the sample. AFM images of single structures could be taken before and after the observation of the optical response.

9.1.1 Localized Emission Response

When the gold nanorods were raster scanned with the laser at an excitation power (measured at the entrance plane of the objective lens) of 1 mW, an optical emission response from the two ends of the rod occured. It was recorded using a single APD. As already described in detail in Chapter 5.3, using the stage piezo crystals, a 3D scan of the response could be performed. Just as for the two-photon luminescence point spread function determined in Chapter 5.3.1, the images were spaced in the z-direction (the direction perpendicular to the sample) in 50 nm steps.

Figure 9.2 a) shows the x-y-plane image of the optical response as obtained by a raster scan of a rod of length 1 μm in the focal plane. Figure 9.2 b)

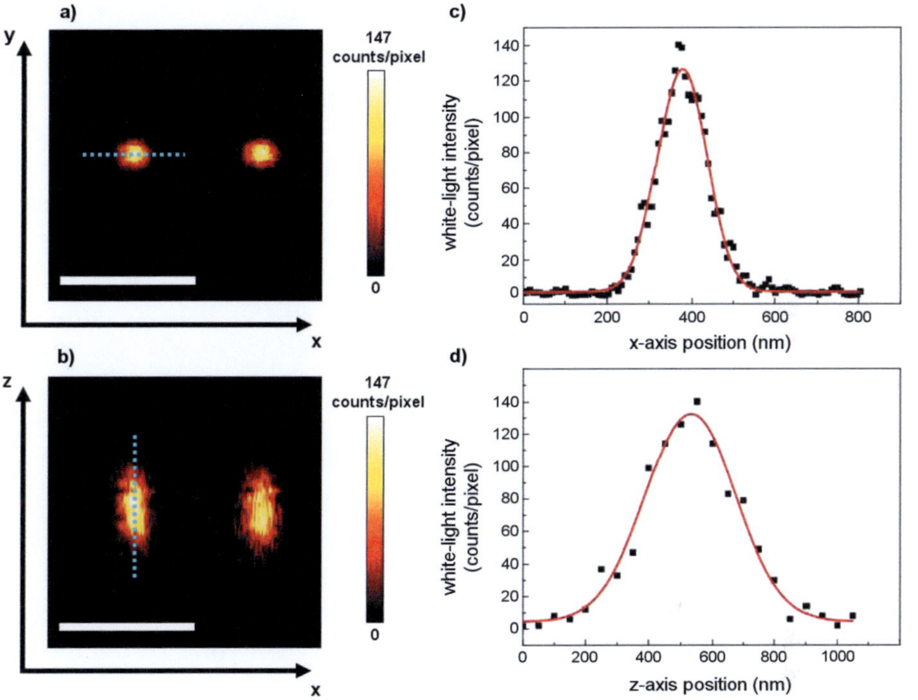

Figure 9.2: The emission signal as collected by a raster scan of a nanorod of length 1 μm. a) Gives the profile as scanned in the x-y-plane, b) depicts a stacked image over the center of the x-z-distribution, where z marks the axis along which the laser light propagates. Pixel integration time was 0.3 μs. Scale bar length is 1 μm. c) and d) give the data points from the light blue dotted line cuts in a) and b) as black dots along the x- and the z-axis respectively; Gaussian fits (red line) are added to the white-light intensity (photon count) data. The FWHM for c) is 138 nm, for d) it is 328 nm. The values for the right spot are very similar.

gives an image of the corresponding white-light distribution in the central x-z-plane as obtained by combination of several such raster scans with the graphic processing software ImageJ (U.S. National Institute of Health). From those images, the interaction cross-sections in these planes can be obtained. Figures 9.2 c) and d) give the measured values as well as a Gaussian fit for both a cut in the x- and the z-direction, respectively. A very narrow distribution with a full-width at half-maximum (FWHM) of 138 nm (cut in x-y plane along x) and 328 nm (cut in x-z plane along z) is obtained.

If the ends of the rods are assumed to be point like detection probes compared to the extent of the laser focus (which is a crude, but still comparably valid assumption), we can estimate the order of the total excitation process via the apparent dimensions of the excitation spot. A Gaussian beam profile in one dimension (in this case, x) can be described by (neglecting units) $C \cdot \exp(-\frac{1}{2}(\frac{x}{\sigma})^2)$, where the FWHM of the profile is described by $2\sigma\sqrt{2\ln(2)}$, x is the position on the x-axis, and C is a normalization factor which only scales the function, but does not influence the FWHM. The FWHM of an n-th order process is then found by taking the n-th power of this profile. Obviously, the FWHM is divided by $\sqrt{n}$ for such an n-th order process. So starting at $0.6098\frac{\lambda}{NA} = 335$ nm (for $\lambda = 800$ nm, NA=1.46), for a FWHM of 138 nm a total excitation process of approximately 6-th order is observed.

However, the logarithmically plotted dependence between excitation power and observed intensity, which is a well-established measure for the order of an excitation process [103], for this measurement still gives a clear indication that a two-photon induced luminescence process is observed – the fitting curve (shown as an inset in Figure 9.5) yields a slope of 2.15. This apparent discrepancy is best explained with the presence of two effects going hand in hand: The strong localization is caused by electromagnetic field line crowding at the ends of the rod (lightning rod effect [69, 111]), which then subsequently enables a two-photon induced luminescence as a secondary process. Configuration errors in the setup of the investigation can be ruled out since the point spread function of the confocal setup with two-photon excitation observed using a single quantum dot is known to correspond to the typically expected large dimensions (compare Chapter 5.3.1).

The observed optical effect is independent of rod length, as can be seen in Figure 9.3. The rod length was reduced from 1 μm to 500 nm in 100 nm steps without any noticeable effect on the strongly localized spots. They simply start to approach each other, with the FWHM for all spots in the x-y-plane comparable in size to the one shown in 3D for the 1 μm long structure in Figure 9.2. Shorter rod lengths are not considered in this experiment to avoid the

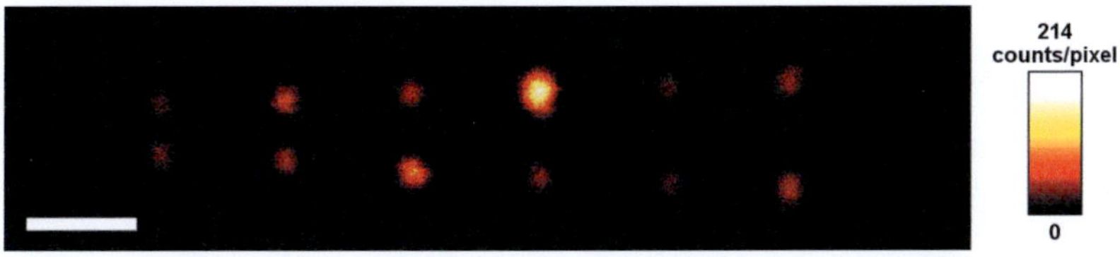

Figure 9.3: The two-photon induced photoluminecence signal for an x-y-plane scan (in the focal plane) of rods of length 500 nm to 1 μm (increasing in 100 nm steps from left to right) as recorded by the single-photon detector. The integration time for each pixel was 0.5 μs. Scale bar length is 1 μm. By trend, the emission profiles at the rod end simply approach each other, with no apparent length influence of the rod. The differences in exact peak shape are attributed to variations in the surface properties of the rods due to imperfections in the electron-beam-lithography process. The fact that the response may differ for one single rod when its two ends are considered is in fact an additional feature suggesting that an effect that is not dependent on the overall length of the rod is observed.

plasmon resonances described in previous chapters. For the long structures, it is therefore justified to assume that a non-resonant lightning-rod effect is observed, in strong contrast to the resonantly generated two-photon induced photoluminescence (TPI-PL) on shorter nanorods or purposely designed nanostructures seen previously. This is also supported by the fact that distinctly more power (factor between 10^2 and 10^3) was required to generate such a non-resonant white-light response from the nanorods compared to the excitation of resonant optical dipole antennas. The amount of white-light generated didn't show any significant dependence on the polarization state of the incoming light (all optical images presented in this work are for an electric field polarization along the long rod axis). This again confirms that a very localized process takes place at the ends of the rod, which is not directly related to possible plasmonic resonances on it.

9.1.2 Morphological Changes

Additionally, an investigation of the geometrical properties of the gold nanorods before and after the excitation using the above-described raster-scanning process was performed. Figure 9.4 a) shows an AFM image of a typical 1 μm nanorod before any excitation has taken place. Figure 9.4 b) shows a typical rod after excitation. Clearly, the structural properties have been changed, producing a dumbbell shape with almost spherical ends. In fact, as can be seen by the cut through the middle axis in Figures 9.4 c) and 9.4 d), the ends seem to have undergone extensive changes in morphology compared to their state before excitation. This transformation happens within a very short time

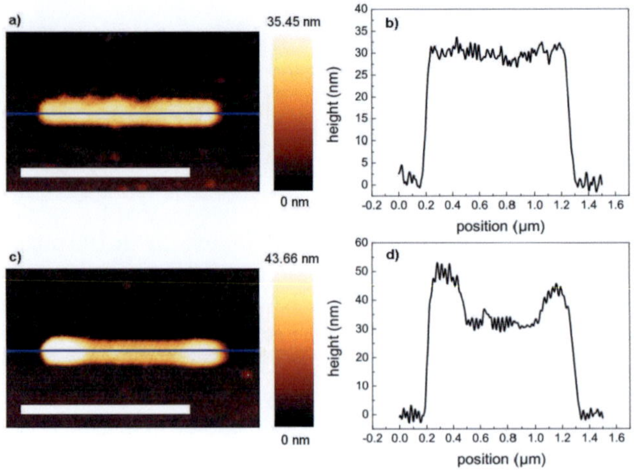

Figure 9.4: a) and c) show AFM images of a 1 μm x 50 nm x 30 nm rod, before and after raster-scanning over the confocal laser volume. The scale bar again indicates 1 μm. b) and d) give the respective cross sections, marked by the blue line in a) and c).

scale. The transformation of the rods was already completed after a first raster-scan had been performed. This is not surprising, however, since for resonantly driven gold nanorods, a transformation was shown to have taken place even after single excitation pulses [201], while for the nanorod investigated here, already after one scan of the rod area, it has already been exposed to ten thousands of pulses. Only the AFM z-dimension is quantitative in the images, since nanostructure AFM images always contain properties of the tip shape as well.

It is well known for smaller gold nanorods that they can be heated by fs laser pulses such that they are completely transformed to nanospheres with similar volume, but new dimensions [201]. This happens due to self-reorganization by minimization of surface energy in the liquid phase which is reached by the high induced powers. It thus seems that the same process applies selectively to the ends of the nanorods where high fields are concentrated due to the lightning rod effect. Since the effect is localized to a very small fraction of the rod and non-resonant, it is, however, difficult to estimate the actual temperature reached by the process, as estimations using bulk parameters would still be valid, but the absorption cross section C_{abs} is essentially unknown.

A rough first order approximation of the temperature that can be reached by the illumination powers that were applied can be given. As mentioned above, the maximum averaged laser power applied (as measured before the objective lens) was 1 mW. For each pulse of approximately 500 fs width (corrected for broadening due to beam-expanding lenses and a dielectric mirror, at 76 MHz repetition rate), a laser fluence I of about $1 \cdot 10^2 \frac{\mathrm{J}}{\mathrm{m}^2}$ is reached, assuming a focal area of π x 200^2 nm^2 and an objective lens transmission efficiency of 0.7 [1] at 800 nm wavelength. The temperature rise ΔT within the sample can be described by

$$\Delta T = \frac{C_{\mathrm{abs}} \cdot I}{m c_p}, \tag{9.1}$$

neglecting a possible latent energy (phase transition), with I the laser fluence, m the mass of the structure, and c_p the bulk specific heat capacity of gold. Since a non-resonant case is considered, it seems to be justified to approximate the absorption cross section by the actual interaction area, which is estimated to be 100 x 50 nm^2 from the AFM image of the transformed rod. Using bulk parameters, the corresponding volume (30 nm height) has a mass m of approximately 3 x 10^{-18} kg (in fact, any absorption cross section corresponding to its volume will yield the same rise in temperature). With $c_p = 129 \frac{\mathrm{J}}{\mathrm{kg\,K}}$ [185], a rough estimate for the rise in temperature is 190 Kelvin, which yields a total temperature of about 480 K. This is still significantly below the bulk melting point of gold (T_{melt}=1337 K).

It can thus be deduced that a field enhancement at the ends of the rods is definitely required to make locally available the necessary energy for a melting effect, which again supports the hypothesis that strongly localized fields are available at the ends of the rods. This is especially valid since the actual temperature reached by pure laser radiation can be assumed to be much lower than in our estimate, since the pulse width of 500 fs is long enough for heat diffusion by phonon-phonon coupling to take place before all energy has been deposited [54].

Interestingly, the selective morphology change at shape singularities of gold nanostructures could open novel nanofabrication schemes for complex plasmonic nanocircuits. The link of optical excitation with relaxation channels that feed radiative as well as non-radiative pathways clearly rises fundamental research interest as well as application oriented questions. Significant interaction of gold nanoparticles with the excitation light has already spawned research concerning laser-induced morphology changes of such nanostructures [101, 183, 54, 201], showing e.g. a transformation from rod shapes to energetically favorable spherical configurations under substantial laser irradiance. The transition rate, the dependence on the photon flux density of the laser excitation pulse as well as the modeling of such processes is a field of active research. Melting effects were found to be most prominent for femtosecond pulses [112] focused on gold nanorods.

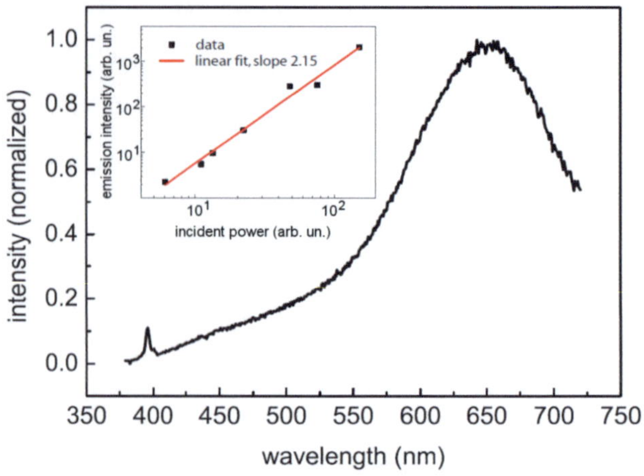

Figure 9.5: Spectrum of the white-light response at the end of one of the gold nanorods. The left peak around 400 nm is due to the second-harmonic generation, while the larger right peak around 650 nm can be explained by interband transitions in gold. The inset shows the emitted white-light intensity (as detected by the APD) versus the incident laser power on a double logarithmic scale. More details are given in the text.

9.1.3 Spectral Properties

The spectral properties of the white-light generation were also investigated. A spectrum measured from one of the ends of a rod of size 1 μm x 50 nm x 30 nm can be seen in Figure 9.5. The spectrum was found to be very similar for all rod lengths under investigation (again from 1 μm to 500 nm) and independent of excitation polarization, with some second-harmonic light generated as well as very large peak around 640 to 670 nm. The second-harmonic generation phenomenon is well-known for localized near-fields [29] and again a strong indicator for a very strongly enhanced field at the rod ends, as well as a measure for its enhancement factor [24]. At first glance, the larger peak around 1.9 eV fits to the transverse mode for a 50 nm wide antenna, but it is significantly broader (FWHM of 0.5 eV) than any gold transverse response earlier, so there must be an additional contribution. When comparing these spectra to published work, it is apparent that the higher peak wavelength also fits to the energy gap between the d-band and the Fermi level at the X-symmetry point of the first Brillouin zone, as already observed for TPI-PL of rods fabricated by colloidal synthesis, where higher order plasmon modes could be observed by near-field microscopy [86]. Thus the peak width and position can also be attributed, at least in part, to material properties in this case. The second interband peak observed by Imura et al. around 550 nm is not present in this experiment. Possibly the difference in fabrication technology between top-down and bottom-up approaches is responsible for the distinctly different behaviour, but there may also be other causes such as additional features induced by the lightning-rod effect.

9.2 Probing of Enhanced Fields Using Quantum Dots

Besides far-field scattering, two-photon luminescence, and near-field approaches, enhanced fields are often also accessed using luminescent substances [47, 173, 104]: Near-fields, which will usually be confined and won't propagate towards the detector, locally increase the far-field response of a luminescent emitter. A similar approach will be presented here to establish additional insight into the fields generated by the non-resonant structures.

The rod structures investigated previously were now additionally covered with a densely packed layer of quantum dots (QD605 carboxyl quantum dots, Invitrogen). To achieve a full covering of the nanostructure array, a drop-casting method was chosen: A small drop of highly concentrated quantum dot solution (concentration 8 μM) was applied to the sample. When the liquid solution had evaporated, a dense layer of quantum dots was present on the sample.

To yield a strong luminescence response from the quantum dots, the sample was then illuminated using a linearly polarized 532 nm peak wavelength CW laser (diode pumped solid state laser DPSS Green 532, HB-Laser), again with a 100 x oil immersion objective lens (NA=1.46). The illumination light was filtered out in the detection channel using a dichroic mirror and an additional wavelength filter (Zeiss microscopy filter set 14). The emitted light could be detected using devices already described previously: a real colour CCD camera, a spectrograph (Acton Spectra Pro 2500i) with an attached Andor iXon EMCCD camera as imaging device, and a single-photon detector.

9.2.1 Real Colour CCD image

When the laser is directly focused onto the sample, there is of course a very strong fluorescence response of the quantum dots at the focus, as seen in Figure 9.6. However, in addition to the focal shape, there are also several bright spots visible outside the focus. These are the positions of the gold nanorods. All responses that can be seen are red, a first indication that only quantum dot

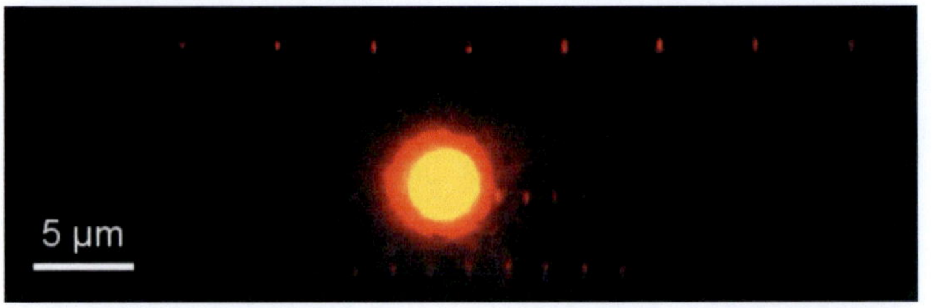

Figure 9.6: Real colour CCD image of a sample containing different non-resonant rods as well as a covering layer of quantum dots. The quantum dot luminescence response is very strong at the focal point (yellow colour due to overexposure), but the luminescence is also clearly enhanced where the gold rods are located.

Figure 9.7: Raster scan image as recorded by the single-photon detector for a quantum dot (QD) excitation with laser light polarized along the long rod axis for rods of toal length 500 nm (left) through 1 μm (right). The emission is clearly enhanced at the rod ends as already seen in the TPL experiment. No explicit rod length dependency is recorded. Only a qualitative emission intensity scale is give, as the exact emission intensity depends significantly on the quantum dot layer concentration at specific points, such that no quantitative value on the emission enhancement can be calculated from the recorded images. The fact that the distribution and emission of the QDs is not the same all over the sample is seen by additional shapes imprinted in the luminescence response.

luminescence is observed and no other nonlinear effects as observed previously are present. The yellowish shape of the focal point is due to an overexposure of the camera in that area to make available the weaker rod responses.

9.2.2 Raster Scanning

To obtain more quantitative information, again a raster scan is performed, albeit this time only in two dimensions. The laser power is reduced significantly to 20 μW in front of the objective lens to allow for a weak, but slightly larger than dark count noise response from the pure quantum dot layer without any metallic structures present on the single photon detectors. The sample is then moved using the piezo stage, and the optical response for each point is collected.

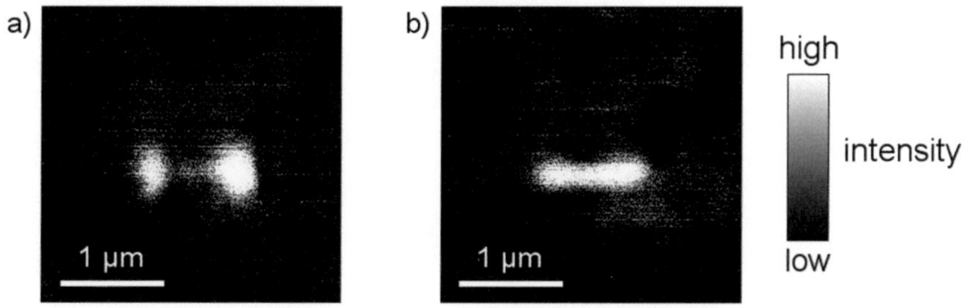

Figure 9.8: Raster scan for a single rod of length 1 μm. In a), excitation polarization is along the long rod axis, in b), along the short in-plane axis. Clearly, the emission pattern from the quantum dots and thus the field distribution near the rod changes significantly.

Previously, the field-enhancement at the rod ends has been attributed to non-resonant lightning-rod effects. If this is true, this effect should be relatively independent of the excitation wavelength. Consequently, for a 532 nm peak wavelength laser illumination, again fields should be concentrated at the rod ends, leading to enhanced quantum dot emission preferably from the rod ends. This response should still be independent of rod length. A raster scan for several such rods is shown in Figure 9.7. Indeed, the luminescence emission is increased at the rod ends exclusively, and there is no significant dependency on total rod length. The size of the spots at the ends is larger, however, and now solely determined by the diffraction limit for the emitted 605 nm wavelength light. This indicates again that a non-resonant field-enhancement is present at the rod ends, providing further evidence for the self-induced nonlinear effects assumed in the previous sections.

Again, a structure of dimensions 1 μm x 50 nm x 30 nm is investigated more closely. Its response is seen in higher resolution in Figure 9.8 a) for an excitation polarization along the long rod axis. Interestingly, this time, a ninety degree change of polarization for the incoming light gives a completely different picture, as seen in Figure 9.8 b). While this may come as a surprise at first, this is not so unexpected when the dimensions of the rod are considered. Despite the resonance red-shift due to the increased quantum dot dielectric constant compared to air, it is still possible that there is some small, but noticeable additional transverse resonance for the now transversely driven rod, as the excitation wavelength has changed significantly towards lower wavelengths. This now gives rise to enhanced fields for each transverse excitation spot, leading to different results compared to what was observed using 800 nm excitation light in the previous section.

Due to CW excitation and thus relatively low excitation peak powers, no

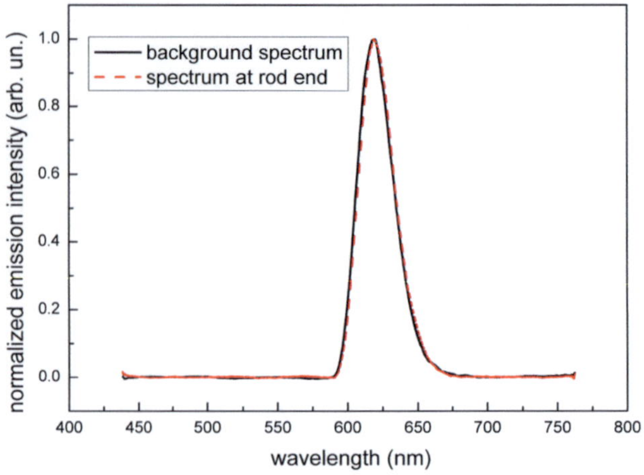

Figure 9.9: Luminescence spectrum for the quantum dots luminescing on the bare substrate (black solid line) and the spectrum including the enhancement at the rod end (red dashed line).

nonlinear conversion effects by the antenna structures are expected, so the emitted light should only come from the quantum dots. To verify this, Figure 9.9 gives a comparison between a spectrum observed from the quantum dots where no metallic structure is near and the spectrum observed at the end of a rod. No noticeable difference is seen in the spectral shape, there is no spectrally distinct response from the rod itself. A more detailed examination of the quantum dot response using lower concentrations would be beneficial, as here the background contribution of quantum dots not influenced by the localized fields is still very high. Also, the use of a pulsed laser instead of CW excitation would enable the examination of luminescence lifetimes.

9.3 Conclusions

Summarizing this chapter, the 3D spatial spread of the white-light system response of gold nanorods upon non-resonant laser illumination was thoroughly investigated. Notably, the 3D optical system response was found to be well beyond the conventionally expected $\lambda/3$-limit for two-photon microscopy. The interpretation of the two-photon induced luminescence as a secondary process which follows the field line crowding at the rod ends due to the lightning rod effect, which determines the spatial extent of the white-light response, enables further studies on light-metal interaction. Interestingly, the localized fields also had distinct effects on structure morphology: Local shape transformations were observed. These transformations are attributed to local heating effects, leading to self-reorganization. The emission spectrum from the 800 nm laser excitation lead to second-harmonic generation properties, as well as a broadband response most likely due to bulk material properties. It was also shown that the very simple method of covering a sample of metallic nanostructures with a layer of quantum dots can lead to additional knowledge about the properties of localized fields very near to the structure. A very limited number of quantum dots near such rods, positioned at specific places, would lead to substantial additional insight. Investigations about single quantum dots as local probes for nanoantennas are currently also performed at the Light Technology Institute.

Chapter 10

Summary and Outlook

In this thesis, extensive investigations of linear and nonlinear optical effects in two-arm and single-arm nanoantenna structures have been described and several proof-of-principle experiments that will facilitate future applications of optical antennas have been carried out.

It was shown that both gold and aluminum antennas operating at optical frequencies can be fabricated reproducibly on a glass substrate (additionally covered with ITO for gold structures) using electron beam lithography exclusively. A powerful characterization setup was developed enabling several characterization schemes, most importantly dark-field microscopy, confocal microscopy, and time-correlated single photon counting. Using the latter, two-photon excitation induced antibunching of quantum dots was demonstrated.

Applying the setup to investigate the optical properties of nanoantennas, far-field scattering investigations of gold two-arm antenna structures coupled by a small gap of 20 nm yielded an eightfold increase in scattering intensity, which shows gap antennas to be efficient scatterers at optical frequencies. Their scattering spectra are also signi-ficantly red-shifted due to the coupling effects compared to single antenna arms. Aluminum two-arm and single-arm antennas were investigated extensively as well. They were found to be much closer to classical RF antennas in terms of their resonance wavelength, with two-arm coupling also taking place via the antenna gap. They are much more broadband than their gold counterparts, which can be beneficial e.g. for the enhancement of solar cells or light-emitting devices.

Additionally, two-photon luminescence of gold two-arm and single-arm antennas provided valuable insight into relaxation channels of nanostructures under pulsed laser excitation. Gold antennas show a response that is dependent in intensity on how resonant the structure is for the excitation wavelength. On the other hand, their spectral response under two-photon excitation conditions can be entirely shaped by particle eigenmodes. This can find useful applications where frequency conversion effects are desired, as any wavelength can be achieved with a proper antenna design. This is especially true as it is demonstrated that this effect does not need to be polarization conserving, which is an unexpected feature for an antenna. For aluminum antennas, a similar response can be observed, but losses are seemingly strongly enhanced, such that melting of antennas can occur quickly and renders spectral observations more challenging. Aluminum antenna luminescence peaks can be observed within the interband transition region of the material, which requires studies on relaxation channels available to nanostructures of that material. Investigations with additional excitation wavelengths and detection filters might provide valuable further insights into this behaviour.

A simple, but efficient apertureless scanning near-field method was introduced to gain experimental access to the near-field concentration of radiation for gold two-arm nanoantennas. While the technique still needs to be optimized further (for example by the use of specifically fabricated tips to yield an optimal scattering and enhancement effect), it provides valuable insight into antenna near-field behaviour and gives a clear indication that a strongly enhanced electromagnetic field is present in the antenna gap.

Finally, it was shown with an extensive study that localized near-fields can also occur near non-resonant structures via the so-called lightning rod effect, and that such a localization of fields can even be observed in the far-field with a resolution below the diffraction limit. It was also shown that this field localization can be probed by luminescent entities like quantum dots, as their emission is significantly enhanced at the specific points of high field concentration.

The results in this work may find numerous applications in the areas already outlined in Chapter 2.6. Three additional ideas are currently under investigation at the Karlsruhe Institute of Technology (KIT): One possible application of the linear wire optical antennas presented here is the improvement of single-photon detectors via field concentration towards a sensitive detection material, for example a superconductor. Superconducting nanowire single-photon detectors could be improved via antennas working in the optical or near-infrared regime. A first attempt to combine aluminum antennas with a niobium-nitride bridge has already been performed and investigations are ongoing in a cooperation with the Institute of Micro- und Nanoelectronic Systems (Prof. Dr. M. Siegel).

As antibunching can be observed from single quantum dots under two-photon excitation, antennas operating in the red wavelength regime might also be beneficially combined with such structures for excitation, generating efficient single photon sources. Investigations on the manipulation of such structures using an AFM to push them in the gap of two-arm nanoantennas are currently carried out at the Light Technology Institute. Using four-arm antennas, even an enhancement of both the excitation and the emission properties of quantum dots can be foreseen.

Finally, optical antennas may also play a crucial role in enhancing the performance of so-called optical rectennas, devices specifically designed to directly convert electromagnetic radiation into a DC current. The enhanced fields in the gap should help to facilitate the use of efficient rectification diodes. These possible applications show that optical antennas will continue to inspire researchers for the years to come.

Abbreviations and Symbols

Symbol	Explanation
1D, 2D, 3D	one-, two- and three-dimensional
2PE	two-photon excitation
a-SNOM	apertureless scattering near-field optical microscopy
AC	amplified current
AFM	atomic force microscope, atomic force microscopy
APD	avalanche photodiode
arb. un.	arbitrary units
CCD	charge-coupled device
CW	continuous-wave
DC	direct current
DPSS	diode-pumped solid-state
EBL	electron beam lithography
EELS	electron energy loss spectroscopy
EMCCD	electron multiplying charge-coupled device
EMF	electromagnetic field
FDTD	finite-difference time-domain
FLIM	fluorescence lifetime imaging
FWHM	full width at half-maximum
IEEE	Institute of Electrical and Electronics Engineers
IMS	Institute of Micro- and Nanoelectronic Systems
IR	infrared
ITO	indium tin oxide

LTI	Light Technology Institute
NA	numerical aperture
OD	optical density
PL	photoluminescence
PML	perfectly matched layer
PMMA	polymethyl methacrylate
PSF	point spread function
QD	quantum dot
RF	radio frequency
SEM	scanning electron beam microscopy
SNOM	scattering near-field optical microscopy
SNR	signal-to-noise ratio
TAO	tip-assisted optics
TCSPC	time-correlated single photon counting
Ti:Sa	Titanium-Sapphire
TPI	two-photon induced
TPL	two-photon luminescence
UV	ultraviolet
$\mathbf{A}$	vector potential
A_e	antenna effective area
$\mathbf{B}$	magnetic induction
c	speed of light
C_abs	absorption cross section
c_p	bulk specific heat capacity
C_sca	scattering cross section
$\mathbf{D}$	electric displacement field
D	directivity
$\mathbf{E}$	electric field
E_g	band gap energy
$\mathbf{e}_x$	unit vector in x direction
$g^2(\tau_\mathrm{inter})$	correlation function

G	gain
$\mathbf{H}$	magnetic field
H	antenna height
I	current amplitude; optical intensity; laser fluence
$\mathrm{Im}\{x\}$	imaginary part of x
$\mathbf{j}$	electric current density
j	imaginary unit
k	decay rate
L	length
L_i	geometrical factor
$\mathbf{m}$	dipole moment
m	relative complex dielectric function
m_eff	effective mass
n	index of refraction
∇	del operator
$\mathbf{P}$	polarization
P	power
p_lf	polarization loss factor
Q	quality factor
$\mathbf{r}$	position
R	Ohmic resistance; radius; photoluminescence rate
$\mathrm{Re}\{x\}$	real part of x
$\mathbf{S}$	Poynting vector
T	temperature
t	time
t_r	radiative lifetime
U	radiation intensity
V	volume; voltage
W	antenna width
x	size parameter
Z	complex impedance
Z_0	free space impedance

α	polarizability
β	propagation constant
γ	damping factor
Δ	antenna gap size
δ	skin depth
$\delta(x)$	delta distribution evaluated at x
ϵ	dielectric function
ζ	Riccati-Bessel function (together with ψ)
η	radiation efficiency; quantum efficiency
λ	wavelength
μ	magnetic permeability
ρ	free electric charge density
ς	spring constant
τ	(apparent) lifetime
τ_{inter}	shifting time for correlation
Φ	scalar potential
ϕ_{ref}	phase shift
χ_{e}	electric susceptibility
ψ	Riccati-Bessel function (together with ζ)
ω	angular frequency

Acknowledgements

I would like to extend my gratitude to Professor Dr. Uli Lemmer for his guidance through the PhD program and for allowing me to pursue the interesting effects encountered with optical antennas at the Light Technology Institute (LTI).

I am greatly indebted to my supervisor Dr. habil. Hans-Jürgen Eisler, who leads the *DFG Heisenberg Nanoscale Science Group* at the LTI in which this work was carried out. The results presented here could not have been achieved without his unceasing support and enthusiasm for my research. Thank you for all scientific discussions, help in the lab, group dinners, and last but not least the fact that you brought optical antenna research to the LTI!

For the support I received at the Institute of Micro- and Nanoelectronic Systems (IMS) to fabricate the nanostructures used throughout this work, I am very grateful to Prof. Dr. Michael Siegel and Dr. Konstantin Ilin. None of the antennas presented here could have existed without the support my students and I received from them. I also thank Hansjürgen Wermund for performing the gold evaporation for the antennas.

I would also like to thank Prof. Dr. Gert Trommer, Prof. Dr. Klaus Müller-Glaser, and Prof. Dr. Ingmar Kallfass for being part of my examination board.

I am especially grateful to all past and current members of the Nanoscale Science group, especially Birgit Rudat, Carola Moosmann, and Katja Dopf. It was great to work with you! I also thank Andreas Schell for starting the setup of the confocal microscopy and TCSPC system with me.

A thank-you note also goes to all of my students – Gabor Varga, Patrick Schwab, Fabian Herberger, Ekkehart Schmidt, and William Truong. All your effort is greatly appreciated and led to wonderful results! I also thank the students who greatly supported my work as HiWis: Daniel Gandyra, Thorsten Beuth, and Xin Liu.

A special thank-you goes to Sönke Klinkhammer for organizing the *Festkörperelektronik* exercise course with me, as well as Birgit Rudat for helping me with the *Optics on the Nanoscale* laboratory course.

I also thank all colleagues at the LTI who helped me whenever I needed scientific advice – especially Felix Glöckler, Ulf Geyer, Klaus Huska, Yousef Nazirizadeh, and Sebastian Valouch. For support with technical matters, I also thank Felix Geislhöringer and all members of the LTI workshop, especially Mario Sütsch. Astrid Dittrich and Claudia Holeisen made sure that I was never lost in the mist and midst of bureaucracy by running the probably most efficient secretariat at the KIT, and Siegfried Kettlitz and Carsten Eschenbaum solved all IT problems. Finally, Christian Kayser and Torsten Feldmann made sure the cleanroom facilities at the LTI were always in a good shape. Thank you!

Thank you to the *Karlsruhe School of Optics and Photonics* for all the events offered and the scholarship I received during the first year of my PhD. I also thank the *DFG* for funding the lion's share of this research project.

To Birgit, Katja, Sönke, Ziyao, and Caro – thank you for making my time in room 212 a wonderful one! Special honours go to Boris, Julian, Carola, Florian, and Tobias for making other rooms almost as comfortable.

I also thank all members of the LTI jogging group, especially Boris and Carola, who made sure I didn't move around in the lab exclusively. A thank-you also goes to the coffee group after lunch and all organizers and participants of LTI social events.

For the proofreading of this manuscript, I especially thank Hans, Benedikt, Carola, Katja, and Birgit.

For unlimited support and help I thank my family, particularly my parents and my brother, who are always there when needed.

The most heartfelt thank-you of all goes to my girlfriend Antonia. Thank you for always believing in me.

List of Figures

List of Tables

Bibliography

[1] Carl Zeiss AG. Manufacturer data provided by Zeiss.

[2] J. Aizpurua, G.W. Bryant, L.J. Richter, F.J. García de Abajo, B.K. Kelley, and T. Mallouk. Optical properties of coupled metallic nanorods for field-enhancement spectroscopy. *Phys. Rev.B*, 71:254320, 2005.

[3] M.G. Albrecht and J.A. Creighton. Anomalously intense Raman spectra of pyridine at a silver electrode. *J. Am. Chem. Soc.*, 99:5215, 1977.

[4] A.P. Alivisatos. Semiconductor clusters, nanocrystals, and quantum dots. *Science*, 271:933, 1996.

[5] A. Alu and N. Engheta. Tuning the scattering response of optical nanoantennas with nanocircuit loads. *Nature Photon.*, 2:307, 2008.

[6] P. Anikeeva, J.E. Halpert, M.G. Bawendi, and V. Bulovic. Quantum dot light-emitting devices with electroluminescence tunable over the entire visible spectrum. *Nano Lett.*, 9:2532, 2009.

[7] Antenna Standards Committee of the IEEE Antennas and Propagation Society, editor. *IEEE standard definitions of terms for antennas - IEEE Std 145-1993*. IEEE, 1993.

[8] J. Azoulay, A. Débarre, A. Richard, and P. Tchénio. Quenching and enhancement of single-molecule fluorescence under metallic and dielectric tips. *Europhys. Lett.*, 51:374, 2000.

[9] C.A. Balanis. Antenna theory: A review. *Proc. IEEE*, 80:7, 1992.

[10] C.A. Balanis. *Antenna Theory - Analysis and Design, Second Edition*. John Wiley and Sons, Inc., 1997.

[11] D.J. Barber and I.C. Freestone. An investigation of the origin of the colour of the Lycurgus Cup by analytical transmission electron microscopy. *Archaeometry*, 32:33, 1990.

[12] W. Becker. *The bh TCSPC Handbook*. Becker & Hickl GmbH, 2010.

[13] J.S. Belrose. Fessenden and Marconi: Their differing technologies and transatlantic experiments during the first decade of this century. In *International Conference on 100 Years of Radio*, 1995.

[14] B. Berland. Photovoltaic technologies beyond the horizon: Optical rectenna solar cell. Technical report, National Renewable Energy Laboratory, Golden, Colorado, USA, 2003.

[15] P. Bharadwaj, B. Deutsch, and L. Novotny. Optical antennas. *Advances in Optics and Photonics*, 1:438, 2009.

[16] P. Bharadwaj and L. Novotny. Spectral dependence of single molecule fluorescence enhancement. *Opt. Exp.*, 15:14266, 2007.

[17] P. Biagioni, J.N. Farahani, P. Mühlschlegel, H.J. Eisler, D.W. Pohl, and B. Hecht. A simple method for producing flattened atomic force microscopy tips. *Rev. Sci. Instrum.*, 79:016103, 2008.

[18] P. Biagioni, J.-S. Huang, and B. Hecht. Nanoantennas for visible and infrared radiation. *arXiv:1103.1568v1*, 2011.

[19] P. Biagioni, M. Savoini, J.-S. Huang, L. Duò, M. Finazzi, and B. Hecht. Near-field polarization shaping by a near-resonant plasmonic cross antenna. *Phys. Rev. B*, 80:153409, 2009.

[20] R.X. Bian, R.C. Dunn, X.S. Xie, and P.T. Leung. Single molecule emission characteristics in near-field microscopy. *Phys. Rev. Lett.*, 75:4772, 1995.

[21] C.F. Bohren and D.R. Huffman. *Absorption and Scattering of Light by Small Particles*. WILEY-VCH, 2004.

[22] M. Bosman, V.J. Keast, M. Watanabe, A.I. Maaroof, and M.B. Cortie. Mapping surface plasmons at the nanometre scale with an electron beam. *Nanotechnology*, 18:165505, 2007.

[23] A. Bouhelier, R. Bachelot, G. Lerondel, S. Kostcheev, P. Royer, and G.P. Wiederrecht. Surface plasmon characteristics of tunable photoluminescence in single gold nanorods. *Phys. Rev. Lett.*, 95:267405–(1–4), 2005.

[24] A. Bouhelier, M. Beversluis, A. Hartschuh, and L. Novotny. Near-field second harmonic generation induced by local field enhancement. *Phys. Rev. Lett.*, 90:013903, 2003.

[25] A. Bouhelier, M.R. Beversluis, and L. Novotny. Characterization of nanoplasmonic structures by locally excited photoluminescence. *Appl. Phys. Lett.*, 83:5041–5043, 2003.

[26] G.T. Boyd, Th. Rasing, J.R.R. Leite, and Y.R. Shen. Local-field enhancement on rough surfaces of metals, semimetals, and semiconductors with the use of optical second-harmonic generation. *Phys. Rev. B*, 30:519, 1984.

[27] G.T. Boyd, Z.H. Yu, and Y.R. Shen. Photoinduced luminescence from the noble metals and its enhancement on roughened surfaces. *Phys. Rev. B*, 33:7923, 1986.

[28] R.W. Boyd. *Nonlinear Optics, Second Edition.* Academic Press, 2003.

[29] S.I. Bozhevolnyi, J. Beermann, and V. Coello. Direct observation of localized second-harmonic enhancement in random metal nanostructures. *Phys. Rev. Lett.*, 90:197403, 2003.

[30] L.E. Brus. Electron-electron and electron-hole interactions in small semiconductor crystallites: The size dependence of the lowest excited electronic state. *J. Chem*, 80:4403, 1984.

[31] T.J. Bukowski and J.H. Simmons. Quantum dot research: Current state and future prospects. *Critical Reviews in Solid State and Materials Sciences*, 27:119, 2002.

[32] M. Burresi, D. van Oosten, T. Kampfrath, H. Schoenmaker, R. Heideman, A. Leinse, and L. Kuipers. Probing the magnetic field of light at optical frequencies. *Science*, 326:550, 2009.

[33] K. R. Catchpole and A. Polman. Plasmonic solar cells. *Opt. Exp.*, 16:21793, 2008.

[34] C.K. Chen, T.F. Heinz, D. Ricard, and Y.R. Shen. Surface-enhanced second-harmonic generation and Raman scattering. *Phys. Rev. B*, 27:1965, 1983.

[35] J. Chen and X.-S. Ge. An improvement on the prediction of optical constants and radiative properties by introducing an expression for the damping frequency in Drude model. *Int. J. of Thermophysics*, 21:269, 2000.

[36] S.L. Chuang. *Physics of Optoelectronic Devices*. Wiley-Interscience, 1995.

[37] R.E. Collin. *Antennas and Radiowave Propagation*. McGraw-Hill, 1985.

[38] V.L. Colvin, M.C. Schlamp, and A.P. Alivisatos. Light-emitting diodes made from cadmium selenide nanocrystals and a semiconducting polymer. *Nature*, 370:354, 1994.

[39] William H. Cropper. *Great Physicists*. Oxford University Press, 2001.

[40] E. Cubukcu, E.A. Kort, K.B. Crozier, and F. Capasso. Plasmonic laser antenna. *Appl. Phys. Lett.*, 89:093120, 2006.

[41] A. Curto, T. Taminiau, G. Volpe, M. Kreuzer, R. Quidant, and N.F. van Hulst. A nano-optical Yagi-Uda antenna driven by single emitters. In *European Conference on Antennas and Propagation 2010, 12-16 April 2010, Barcelona*, 2010.

[42] A.G. Curto, G. Volpe, T.H. Taminiau, M.P. Kreuzer, R. Quidant, and N.F. van Hulst. Unidirectional emission of a quantum dot coupled to a nanoantenna. *Science*, 329:930, 2010.

[43] B.O. Dabbousi, J. Rodriguez-Viejo, F.V. Mikulec, J.R. Heine, H. Mattousi, R. Ober, K.F. Jensen, and M.G. Bawendi. (CdSe)ZnS core-shell quantum dots: Synthesis and characterization of a size series of highly luminescent nanocrystallites. *J. Phys. Chem. B*, 101:9463, 1997.

[44] C. Dahmen, B. Schmidt, and G. von Plessen. Radiation damping in metal nanoparticle pairs. *Nano Letters*, 7:318–322, 2007.

[45] M. Danckwerts and L. Novotny. Optical frequency mixing at coupled gold nanoparticles. *Phys. Rev. Lett.*, 98:026104, 2007.

[46] W. Denk, J. Strickler, and W.W. Webb. Two-photon laser scanning fluorescence microscopy. *Science*, 248:73, 1990.

[47] H. Ditlbacher, J.R. Krenn, N. Felidj, B. Lamprecht, G. Schider, M. Salerno, A. Leitner, and F.R. Aussenegg. Fluorescence imaging of surface plasmon fields. *Appl. Phys. Lett.*, 80:404, 2002.

[48] J. Dorfmüller, R. Vogelgesang, W. Khunsin, C. Rockstuhl, C. Etrich, and K. Kern. Plasmonic nanowire antennas: Experiment, simulation, and theory. *Nano Lett.*, 10:3596, 2010.

[49] J. Dorfmüller, R. Vogelgesang, R.T. Weitz, C. Rockstuhl, C. Etrich, T. Pertsch, F. Lederer, and K. Kern. Fabry-Pérot resonances in one-dimensional plasmonic nanostructures. *Nano Lett.*, 9:2372, 2009.

[50] D. Dregeley, R. Taubert, J. Dorfmüller, R. Vogelgesang, K. Kern, and H. Giessen. 3D optical Yagi-Uda nanoantenna array. *Nature Comm.*, 2:267, 2011.

[51] E. Dulkeith, T. Niedereichholz, T.A. Klar, J. Feldmann, G. von Plessen, D.I. Gittins, K.S. Mayya, and F. Caruso. Plasmon emission in photoexcited gold nanoparticles. *Phys. Rev. B*, 70:205424, 2004.

[52] A. Efros and N. Rosen. The electronic structure of semiconductor nanocrystals. *Annu. Rev. Mater. Sci.*, 30:475, 2000.

[53] H. Ehrenreich, H.R. Philipp, and B. Segall. Optical properties of aluminum. *Phys. Rev.*, 132:1918, 1963.

[54] O. Ekici, R.K. Harrison, N.J. Durr, D.S. Eversole, M. Lee, and A. Ben-Yakar. Thermal analysis of gold nanorods heated with femtosecond laser pulses. *J. Phys. D: Appl. Phys.*, 41:185501, 2008.

[55] A.I. Ekimov, F. Hache, M.C. Schanneklein, D. Ricard, C. Flytzanis, I.A. Kudryavtsev, T.V. Yazeva, A.V. Rodina, and A. Efros. Absorption and intensity-dependent photoluminescence measurements on CdSe quantum dots: assignment of the first electronic transistions. *J. Opt. Soc. Am. B*, 10:100, 1993.

[56] N. Engheta. Circuits with light at nanoscales: Optical nanocircuits inspired by metamaterials. *Science*, 317:1698–1702, 2007.

[57] N. Engheta, A. Salandrino, and A. Alu. Circuit elements at optical frequencies: Nanoinductors, nanocpacitors, and nanoresistors. *Phys. Rev. Lett.*, 95:095504, 2005.

[58] J.N. Farahani, H.-J. Eisler, D.W. Pohl, M. Pavius, P. Flückiger, P. Gasser, and B. Hecht. Bow-tie optical antenna probes for single-emitter scanning near-field optical microscopy. *Nanotechnology*, 18:125506, 2007.

[59] J.N. Farahani, D.W. Pohl, H.-J. Eisler, and B. Hecht. Single quantum dot coupled to a scanning optical antenna: A tunable superemitter. *Phys. Rev. Lett*, 95:017402, 2005.

[60] S. Felekyan, R. Kühnemuth, V. Kudryavtsev, C. Sandhagen, W. Becker, and C.A.M. Seidel. Full correlation from picoseconds to seconds by time-resolved and time-correlated single photon detection. *Rev. Sci. Instrum.*, 76:083104, 2005.

[61] U.C. Fischer and D. Pohl. Observation on single-particle plasmons by near-field optical microscopy. *Phys. Rev. Lett.*, 62:458, 1989.

[62] B. Fisher, J.M. Caruge, D. Zehnder, and M. Bawendi. Room-temperature ordered photon emission from multiexciton states in single CdSe core-shell nanocrystals. *Phys. Rev. Lett.*, 94:087403, 2005.

[63] M. Fleischmann, P. J. Hendra, and A.J. McQuillian. Raman spectra of pyridine adsorbed at a silver electrode. *Chem. Phys. Lett.*, 26:163, 1974.

[64] V. Formenko and D.J. Nesbitt. Solution control of radiative and non-radiative lifetimes: A novel contribution to quantum dot blinking suppression. *Nano Lett.*, 8:287, 2008.

[65] I. Freestone, N. Meeks, M. Sax, and C. Higgitt. The Lycurgus Cup - a Roman nanotechnology. *Gold Bulletin*, 40:270, 2007.

[66] A. M. Funston, C. Novo, T. J. Davis, and P. Mulvaney. Plasmon coupling of gold nanorods at short distances and in different geometries. *Nano Lett.*, 9:1651, 2009.

[67] F.J. García de Abajo. Optical excitations in electron microscopy. *Rev. Mod. Phys.*, 82:209, 2010.

[68] M.F. Garcia-Parajo. Optical antennas focus in on biology. *Nature Photonics*, 2:201–203, 2008.

[69] J.I. Gersten. The effect of surface roughness on surface enhanced Raman scattering. *J. Chem. Phys.*, 72:5779, 1980.

[70] P. Ghenuche, S. Cherukulappurath, and R. Quidant. Mode mapping of plasmonic stars using TPL microscopy. *New J. of Phys.*, 10:105013, 2008.

[71] P. Ghenuche, S. Cherukulappurath, T.H. Taminiau, N.F. van Hulst, and R. Quidant. Spectroscopic mode mapping of resonant plasmon nanoantennas. *Phys. Rev. Lett.*, 101:116805, 2008.

[72] S.K. Ghosh and T. Pal. Interparticle coupling effects on the surface plasmon resonance of gold nanoparticles: From theory to applications. *Chem. Rev.*, 107:4797, 2007.

[73] M. Göppert-Mayer. Über Elementarakte mit zwei Quantensprüngen. *Ann. Phys.*, 403:273, 1931.

[74] R. Grober, R. Schoelkopf, and D. Prober. Optical antenna: Towards a unity efficiency near-field optical probe. *Appl. Phys. Lett.*, 70:1354, 1997.

[75] I. Gur, N.A. Fromer, M.L. Geier, and A.P. Alivisatos. Air-stable all-inorganic nanocrystal solar cells processed from solution. *Science*, 310:462, 2005.

[76] R. Hanbury Brown and R.Q. Twiss. Correlation between photons in two coherent beams of light. *Nature*, 177:27, 1956.

[77] R. Hanbury Brown and R.Q. Twiss. The question of correlation between photons in coherent light rays. *Nature*, 178:1447, 1956.

[78] R. Hanbury Brown and R.Q. Twiss. A test of a new type of stellar interferometer on Sirius. *Nature*, 178:1046, 1956.

[79] A. Hartschuh, E.J. Sánchez, X.S. Xie, and L. Novotny. High-resolution near-field Raman microscopy of single-walled carbon nanotubes. *Phys. Rev. Lett.*, 90:095503, 2003.

[80] B. Hecht. Single emitters and optical antennas. *In Tip Enhancement, S. Kawata and V. M. Shalaev, editors, Elsevier, Amsterdam, The Netherlands*, pages 275–307, 2007.

[81] F. Herberger. *Nanomanipulation von Goldkolloiden.* Studienarbeit, Karlsruhe Institute of Technology (KIT), Germany, 2011.

[82] H.F. Hofmann, T. Kosako, and Y. Kadoya. Design parameters for a nano-optical Yagi-Uda antenna. *New J. of Phys.*, 9:217, 2007.

[83] J.S. Huang, J. Kern, P. Geisler, P. Weinmann, M. Kamp, A. Forchel, P. Biagioni, and B. Hecht. Mode imaging and selection in strongly coupled nanoantennas. *Nano Lett.*, 10:2106, 2010.

[84] K. Imura, T. Nagahara, and H. Okamoto. Plasmon mode imaging of single gold nanorods. *J. Am. Chem. Soc.*, 126:12730, 2004.

[85] K. Imura, T. Nagahara, and H. Okamoto. Near-field two-photon-induced photoluminescence from single gold nanorods and imaging of plasmon modes. *J. Phys. Chem. B*, 109:13214, 2005.

[86] K. Imura and H. Okamoto. Properties of photoluminescence from single gold nanorods induced by near-field two-photon excitation. *J. Phys. Chem. C*, 113:11756, 2009.

[87] J.D. Jackson. *Classical Electrodynamics, Third Edition*. John Wiley and Sons, Inc., 1999.

[88] D.L. Jeanmaire and R.P. van Duyne. Surface Raman spectroelectrochemistry: part I. Heterocyclic, aromatic, and aliphatic amines adsorbed on the anodized silver electrode. *J. Electroanal. Chem.*, 84:1, 1977.

[89] P.B. Johnson and R.W. Christy. Optical constants of noble metals. *Phys. Rev. B*, 6:4370, 1972.

[90] H. Kim, C. Xiang, A.G. Güell, R.M. Penner, and E.O. Potma. Tunable two-photon excited luminescence in single gold nanowires fabricated by lithographically patterned nanowire electrodeposition. *J. Phys. Chem. C*, 112:12721, 2008.

[91] S. Kim, B. Fisher, H.-J. Eisler, and M. Bawendi. Type-II quantum dots: CdTe/CdSe(core/shell) and CdSe/ZnTe(core/shell) heterostructures. *J. Am. Chem. Soc.*, 125:11466, 2003.

[92] T.-H. Kim, K.-S. Cho, E.K. Lee, S.J. Lee, J. Chae, J W. Kim, D.H. Kim, J.-Y. Kwon, G. Amaratunga, S.Y. Lee, B.L. Choi, Y. Kuk, J.M. Kim, and K. Kim. Full-colour quantum dot displays fabricated by transfer printing. *Nature Photon.*, 5:176, 2011.

[93] A. Kinkhabwala, Z. Yu, S. Fan, Y. Avlasevich, K. Mullen, and E.W. Moerner. Large single-molecule fluorescence enhancements produced by a bowtie nanoantenna. *Nat. Photon.*, 3:654, 2009.

[94] M.F.G. Klein, H. Jein, P.-J. Jakobs, S. Linden, N. Meinzer, M. Wegener, V. Saile, and M. Kohl. Electron beam lithography of v-shaped silver nanoantennas. *Microelectronic Engineering*, 86:1078, 2009.

[95] V.I. Klimov, A A. Mikhailovsky, D.W. McBranch, C.A. Leatherdale, and M.G. Bawendi. Quantization of multiparticle Auger rates in semiconductor quantum dots. *Science*, 287:1011, 2000.

[96] B. Knoll and F. Keilmann. Near-field probing of vibrational absorption for chemical microscopy. *Nature*, 399:134, 1999.

[97] M. Köllner and J. Wolfrum. How many photons are necessary for fluorescence-lifetime measurements? *Chem. Phys. Lett.*, 200:199, 1992.

[98] T. Kosako, Y. Kadoya, and H.F. Hofmann. Directional control of light by a nano-optical Yagi-Uda antenna. *Nature Photon.*, 4:312, 2010.

[99] U. Kreibig. Hundert Jahre Mie-Theorie. Optische Eigenschaften von Nanopartikeln. *Physik in unserer Zeit*, 39:281, 2008.

[100] S. Kühn, U. Hakanson, L. Rogobete, and V. Sandoghdar. Enhancement of single-molecule fluorescence using a gold nanoparticle as an optical nanoantenna. *Phys. Rev. Lett.*, 97:017402, 2006.

[101] H. Kurita, A. Takami, and S. Koda. Size reduction of gold particles in aqueous solution by pulsed laser irradiation. *Appl. Phys. Lett.*, 72:789, 1998.

[102] H. Kuwata, H. Tamarau, K. Esumi, and K. Miyano. Resonant light scattering from metal nanoparticles: Practical analysis beyond rayleigh approximation. *Appl. Phys. Lett.*, 83:4625, 2003.

[103] J.R. Lakowicz. *Principles of Fluorescence Sepctroscopy, Third Edition*. Springer, New York, 2006.

[104] J.R. Lakowicz, M.H. Chowdhury, K. Ray, J. Zhang, Y. Fu, R. Badugu, C.R. Sabanayagam, K. Nowaczyk, H. Szmacinski, K. Aslan, and C.D. Geddes. Plasmon-controlled fluorescence: a new detection technology. *Proc. SPIE*, 6099:609909, 2006.

[105] S. Lal, S.E. Clare, and N.J. Halas. Nanoshell-enabled photothermal cancer therapy: Impending clinical impact. *Accounts of Chemical Research*, 41:1842, 2008.

[106] D.R. Larson, W.R. Zipfel, R.M. Williams, S.W. Clark, M.P. Bruchez, F.W. Wise, and W.W. Webb. Water-soluble quantum dots for multiphoton fluorescence imaging in vivo. *Science*, 300:1434, 2003.

[107] N.N. Ledentsov, D. Bimberg, and Zh.I. Alferov. Progress in epitaxial growth and performance of quantum dot and quantum wire lasers. *J. of Lightw. Tech.*, 26:1540, 2008.

[108] D. Leonard, M. Krishnamurthy, C.M Reaves, S.P. Denbaars, and P.M. Petroff. Direct formation of quantum-sized dots from uniform coherent islands of InGaAs on GaAs surfaces. *Appl. Phys. Lett.*, 63:3203, 1993.

[109] M. Leutenegger, R. Rao, R.A. Leitgeb, and T. Lasser. Fast focus field calculations. *Opt. Exp.*, 14:11277, 2006.

[110] A. Lewis, M. Isaacson, A. Harootunian, and A. Muray. Development of a 500 angstrom spatial resolution light microscope. *Ultramicroscopy*, 13:227, 1984.

[111] P.F. Liao and A. Wokaun. Lightning rod effect in surface enhanced Raman scattering. *J. Chem. Phys.*, 76:751, 1982.

[112] S. Link, C. Burda, B. Nikoobakht, and M.A. El-Sayed. Laser-induced shape changes of colloidal gold nanorods using femtosecond and nanosecond pulses. *J. Phys. Chem. B*, 104:6152, 2000.

[113] M. Lippitz, M.A. van Dijk, and M. Orrit. Third-harmonic generation from single gold nanoparticles. *Nano Letters*, 5:799, 2005.

[114] D. Loss and D.P. DiVincenzo. Quantum computation with quantum dots. *Phys. Rev. A*, 57:120, 1998.

[115] Lumerical. FDTD Solution. http://www.lumerical.com.

[116] S.A. Maier and H.A. Atwater. Plasmonics: Localization and guiding of electromagnetic energy in metal/dielectric structures. *J. of Appl. Phys.*, 98:011101, 2005.

[117] B.D. Mangum, C. Mu, and J.M. Gerton. Resolving single fluorophores within dense ensembles: contrast limits of tip-enhanced fluorescence microscopy. *Opt. Exp.*, 16:6183–6193, 2008.

[118] D.M. Mattox. *The Foundations of Vacuum Coating Technology*. Noyes Publications, 2003.

[119] K.T. McDonald. Radiation in the near zone of a center-fed linear antenna, June 2004. Internet publication as accessed on April 5, 2011 on http://puhep1.princeton.edu/~mcdonald/examples/linearantenna.pdf.

[120] A.D. McFarland, M.A. Young, J.A. Dieringer, and R.P. van Duyne. Wavelength-scanned surface-enhanced Raman excitation spectroscopy. *J. Phys. Chem. B*, 109:11279, 2005.

[121] M. Meier and A. Wokaun. Enhanced fields on large metal particles: dynamic depolarization. *Opt. Lett.*, 8:581, 1983.

[122] G. Messin, J.P. Hermier, E. Giacobino, P. Desbiolles, and M. Dahan. Bunching and antibunching in the fluorescence of semiconductor nanocrystals. *Opt. Lett.*, 26:1891, 2001.

[123] A.A. Milner, K. Zhang, and Y. Prior. Floating tip nanolithography. *Nano Lett.*, 8:2017, 2008.

[124] M. Minsky. *Microscopy Apparatus*. U.S. Patent 3,013,467, 1961.

[125] Photonic Bandgap Fibers & Devices Group MIT. Indium tin oxide (ITO). http://mit-pbg.mit.edu/Pages/ITO.html.

[126] M.B. Mohamed, V. Volkov, S. Link, and M.A. El-Sayed. The 'lightning' gold nanorods: fluorescence enhancement of over a million compared to the gold metal. *Chem. Phys. Lett.*, 317:517, 2000.

[127] A. Mooradian. Photoluminescence of metals. *Phys. Rev. Lett.*, 22:185, 1969.

[128] P. Mühlschlegel, H.-J. Eisler, O.J.F. Martin, B. Hecht, and D.W. Pohl. Resonant optical antennas. *Science*, 308:1607, 2005.

[129] K. Munechika, Y. Chen, A.F. Tillack, A.P. Kulkarni, I. Jen-La Plante, A.M. Munro, and D.S. Ginger. Spectral control of plasmonic emission enhancement from quantum dots near single silver nanoparticles. *Nano Lett.*, 10:2598, 2010.

[130] K.-I. Murata and H. Tanaka. Surface-wetting effects on the liquid-liquid transition of a single-component molecular liquid. *Nature Comm.*, 1:16, 2010.

[131] C.J. Murphy and J.L. Coffer. Quantum dots: A primer. *Appl. Spectroscopy*, 56:16A, 2002.

[132] C.B. Murray, D.J. Norris, and M.G. Bawendi. Synthesis and characterization of nearly monodisperse CdE (E = sulfur, selenium, tellurium) semiconductor nanocrystallites. *J. Am. Chem. Soc.*, 115:8706, 1993.

[133] O.L. Muskens, V. Giannini, J.A. Sánchez, and J. Gómez-Rivas. Optical scattering resonances of single and coupled dimer plasmonic nanoantennas. *Opt. Exp.*, 15:17736, 2007.

[134] J. Nelayah, M. Kociak, O. Stephan, F.J. García de Abajo, M. Tence, L. Henrard, D. Taverna, I. Pastoriza-Santos, L.M. Liz-Marzan, and C. Colliex. Mapping surface plasmons on a single metallic nanoparticle. *Nature Phys.*, 3:348, 2007.

[135] M. Nirmal, B.O. Dabbousi, M.G. Bawendi, J.J. Macklin, J.K. Trautman, T.D. Harris, and L.E. Brus. Fluorescence intermittency in single cadmium selenide nanocrystals. *Nature*, 383:802, 1996.

[136] P. Nordlander, C. Oubre, E. Prodan, K. Li, and M.I. Stockman. Plasmon hybridization in nanoparticle dimers. *Nano Letters*, 4:899, 2004.

[137] L. Novotny. Effective wavelength scaling for optical antennas. *Phys. Rev. Lett.*, 98:266802, 2007.

[138] L. Novotny, R.X. Brian, and X.S. Xie. Theory of nanometric optical tweezers. *Phys. Rev. Lett.*, 79:645, 1997.

[139] L. Novotny and B. Hecht. *Principles of Nano-Optics*. Cambridge University Press, 2006.

[140] L. Novotny and S.J. Stranick. Near-field optical microscopy and spectroscopy with pointed probes. *Annu. Rev. Phys. Chem.*, 57:303, 2006.

[141] L. Novotny and N. van Hulst. Antennas for light. *Nature Phot.*, 5:83, 2011.

[142] E. Ögüt, G. Kizilitas, and K. Sendur. Circularly polarized localized near-field radiation at the nanoscale. *Appl. Phys. B*, 99:67, 2010.

[143] R.L. Olmon, P.M. Krenz, A.C. Jones, G.D. Boreman, and M.B. Raschke. Near-field imaging of optical antenna modes in the mid-infrared. *Opt. Exp.*, 16:20295, 2008.

[144] M.A. Ordal, R.J. Bell, R.W. Alexander Jr., L.L. Long, and M.R. Querry. Optical properties of fourteen metals in the infrared and far infrared: Al, Co, Cu, Au, Fe, Pb, Mo, Ni, Pd, Pt, Ag, Ti, V, and W. *Appl. Opt.*, 24:4493, 1985.

[145] E.D. Palik, editor. *Handbook of Optical Constants of Solids*. Academic Press, 1984.

[146] H. Paul. Photon antibunching. *Rev. Mod. Phys.*, 54:1061, 1982.

[147] J.B. Pawley, editor. *Handbook of Biological Confocal Microscopy, Third Edition*. Springer, 2006.

[148] M.L. Pedano, S. Li, G.C. Schatz, and C.A. Mirkin. Periodic electric field enhancement along gold rods with nanogaps. *Angew. Chem. Int. Ed.*, 49:78, 2010.

[149] X. Peng, M.C. Schlamp, A.V. Kadavanich, and A.P. Alivisatos. Epitaxial growth of highly luminescent CdSe/CdS core/shell nanocrystals with photostability and electronic accessibility. *J. Am. Chem. Soc.*, 119:7019, 1997.

[150] D. Pohl. Near field optics seen as an antenna problem. *In Near-Field Optics: Principles and Applications, M. Ohtsu and X. Zhu, editors, World Scientific, Singapore*, pages 9–21, 2000.

[151] D. Pohl and L. Novotny. Stacked optical antennas. *Appl. Phys. Lett.*, 98:023111, 2011.

[152] D.W. Pohl, W. Denk, and M. Lanz. Optical stethoscopy: image recording with resolution l/20. *Appl. Phys. Lett.*, 44:651, 1984.

[153] E. Prodan, C. Radloff, N.J. Halas, and P. Nordlander. A hybridization model for the plasmon response of complex nanostructures. *Science*, 302:419, 2003.

[154] J. Renger, R. Quidant, N. van Hulst, and L. Novotny. Surface-enhanced nonlinear four-wave mixing. *Phys. Rev. Lett.*, 104:046803, 2010.

[155] M. Righini, P. Ghenuche, S. Cherukulappurath, V. Myroshnychenko, F.J. García de Abajo, and R. Quidant. Nano-optical trapping of Rayleigh particles and Escherichia coli bacteria with resonant optical antennas. *Nano Letters*, 9:3387, 2009.

[156] C. Rockstuhl, M.G. Salt, and H.P. Herzig. Analyzing the scattering properties of coupled metallic nanoparticles. *J. Opt. Soc. Am. A*, 21:1761, 2004.

[157] A. Sanchez, C.F. Davis, K.C. Liu, and A. Javan. The MOM tunneling diode: Theoretical estimate of its performance at microwave and infrared frequencies. *J. Appl. Phys.*, 49:5270, 1978.

[158] S. Sapra and D.D. Sarma. Evolution of the electronic structure with size in II-VI semiconductor nanocrystals. *Phys. Rev. B*, 69:125304, 2004.

[159] A.W. Schell, G. Kewes, T. Hanke, A. Leitenstorfer, R. Bratschitsch, O. Benson, and T. Aichele. Single defect centers in diamond nanocrystals as quantum probes for plasmonic nanostructures. *Opt. Exp.*, 19:7914, 2011.

[160] E. Schmidt. *Planare optische Antennen aus Aluminium*. Bachelorarbeit, Karlsruhe Institute of Technology (KIT), 2010.

[161] M. Schnell, A. García-Etxarri, A.J. Huber, K. Crozier, J. Aizpurua, and R. Hillenbrand. Controlling the near-field oscillations of loaded plasmonic nanoantennas. *Nature Phot.*, 3:287, 2009.

[162] P.J. Schuck, D.P. Fromm, A. Sundaramurthy, G.S. Kino, and W.E. Moerner. Improving the mismatch between light and nanoscale objects with gold bowtie nanoantennas. *Phys. Rev. Lett.*, 94:017402, 2005.

[163] L.G. Schulz. The optical constants of silver, gold, copper, and aluminum. I. The absorption coefficient k. *J. Opt. Soc. Am.*, 44:357, 1954.

[164] L.G. Schulz and F.R. Tangherlini. Optical constants of silver, gold, copper, and aluminum. II. The index of refraction n. *J. Opt. Soc. Am.*, 44:362, 1954.

[165] P. M. Schwab. *Herstellung und Charakterisierung optischer Quadrupol-Antennen*. Studienarbeit, Karlsruhe Institute of Technology (KIT), Germany, 2010.

[166] P. M. Schwab. *Optische Antennen aus Aluminium.* Diplomarbeit, Karlsruhe Institute of Technology (KIT), Germany, 2011.

[167] M. Scolari, A. Mews, N. Fu, A. Myaltsin, T. Assmus, K. Balasubramanian, M. Burghard, and K. Kern. Surface enhanced Raman scattering of carbon nanotubes decorated by individual fluorescent gold particles. *J. Phys. Chem. C*, 112:391, 2008.

[168] A.M. Smith, X. Gao, and S. Nie. Quantum dot nanocrystals for in vivo molecular and cellular imaging. *Photochem. and Photobiol.*, 80:377, 2004.

[169] E.J. Smythe, E. Cubukcu, and F. Capasso. Optical properties of surface plasmon resonances of coupled metallic nanorods. *Opt. Exp.*, 15:7439, 2007.

[170] C. Sönnichsen. *Plasmons in metal nanostructures.* PhD thesis, LMU Munich, Germany, 2001.

[171] J. Steidtner and B. Pettinger. Tip-enhanced Raman spectroscopy and microscopy on single dye molecules with 15 nm resolution. *Phys. Rev. Lett.*, 100:236101, 2008.

[172] R.M. Stöckle, Y.D. Suh, V. Deckert, and R. Zenobi. Nanoscale chemical analysis by tip-enhanced Raman spectroscopy. *Chem. Phys. Lett.*, 318:131, 2000.

[173] O. Stranik, H.M. McEvoy, C. McDonagh, and B.D. MacCraith. Plasmonic enhancement of fluorescence for sensor applications. *Sensors and Actuators B*, 107:148, 2005.

[174] E.H. Synge. A suggested method for extending microscopic resolution into the ultra-microscopic region. *Phil. Mag.*, 6:356, 1928.

[175] T. Taminiau, F. Stefani, and N. van Hulst. Optical nanorod antennas modeled as cavities for dipolar emitters: Evolution of sub- and super-radiant modes. *Nano Lett.*, 11:1020, 2011.

[176] T.H. Taminiau, F.D. Stefani, F.B. Segerink, and N.F. van Hulst. Optical antennas direct single-molecule emission. *Nature Phot.*, 2:234, 2008.

[177] W. Truong. *Entwicklung und Aufbau eines Konfokaldunkelfeldmikroskopiesystems unter Verwendung kegelförmiger Linsen.* Bachelorarbeit, Karlsruhe Institute of Technology (KIT), Germany, 2010.

[178] W.G. van Sark, P.L. Frederix, A.A. Bol, H.C. Gerritsen, and A. Meijerink. Blueing, bleaching, and blinking of single CdSe/ZnS quantum dots. *ChemPhysChem*, 3:871, 2002.

[179] G. Varga. *Nahfeldcharakterisierung optischer Antennen*. Diplomarbeit, Karlsruhe Institute of Technology (KIT), Germany, 2009.

[180] C. Vion, P. Spinicelli, L. Coolen, C. Schwob, J.-M. Frigerio, J.-P. Hermier, and A. Maître. Controlled modification of single colloidal CdSe/ZnS nanocrystal fluorescence through interactions with a gold surface. *Opt. Exp.*, 18:7440, 2010.

[181] R. Vogelgesang, J. Dorfmüller, R. Esteban, R.T. Weitz, A. Dmitriev, and K. Kern. Plasmonic nanostructures in aperture-less scanning near-field optical microscopy (aSNOM). *phys. stat. sol. (b)*, 245:2255, 2008.

[182] D.-S. Wang, F.-Y. Hsu, and C.-W. Lin. Surface plasmon effects on two photon luminescence of gold nanorods. *Opt. Exp.*, 17:11350–11359, 2009.

[183] Y. Wang and C. Dellago. Structural and morphological transitions in gold nanorods: A computer simulation study. *J. Phys. Chem. B*, 107:9214, 2003.

[184] D.R. Ward, F. Hüser, F. Pauly, J. Carlos Cuevas, and D. Natelson. Optical rectification and field enhancement in a plasmonic nanogap. *Nature Nanotechn.*, 5:732, 2010.

[185] R.C. Weast and D.R. Lide, editors. *CRC Handbook of Chemistry and Physics*. CRC Press, 85th edition, 2004.

[186] J. Wessel. Surface-enhanced optical microscopy. *J. Opt. Soc. Am. B*, 2:1538, 1985.

[187] R.R. Wiley. *Practical Equipment, Materials, and Processes for Optical Thin Films*. Wiley Optical, 2007.

[188] M.D. Wissert. *Optimierung und Charakterisierung optischer Antennen*. Diplomarbeit, Universität Karlsruhe (TH), Germany, 2008.

[189] M.D. Wissert, K.S. Ilin, M. Siegel, U. Lemmer, and H.-J. Eisler. Coupled nanoantenna plasmon resonance spectra from two-photon laser excitation. *Nano Lett.*, 10:4161, 2010.

[190] M.D. Wissert, K.S. Ilin, M. Siegel, U. Lemmer, and H.-J. Eisler. Highly localized non-linear optical white light response at nanorod ends from non-resonant excitation. *Nanoscale*, 2:1018, 2010.

[191] M.D. Wissert, C. Moosmann, K.S. Ilin, M. Siegel, U. Lemmer, and H.-J. Eisler. Gold nanoantenna resonance diagnostics via transversal particle plasmon luminescence. *Opt. Exp.*, 19:3686, 2011.

[192] M.D. Wissert, B. Rudat, U. Lemmer, and H.-J. Eisler. Quantum dots as single-photon sources: Antibunching via two-photon excitation. *Phys. Rev. B*, 83:113304, 2011.

[193] M.D. Wissert, A.W. Schell, K.S. Ilin, M. Siegel, and H.-J. Eisler. Nanoengineering and characterization of gold dipole-antennas with enhanced integrated scattering properties. *Nanotechnology*, 20:425203, 2009.

[194] A. Wokaun, J. Gordon, and P. Liao. Radiation damping in surface-enhanced Raman scattering. *Phys. Rev. Lett.*, 48:957, 1982.

[195] N. Yu, E. Cubukcu, L. Diehl, M.A. Belkin, K.B. Crozier, F. Capasso, D. Bour, S. Corzine, and G. Höfler. Plasmonic quantum cascade laser antenna. *Appl. Phys. Lett.*, 91:173113, 2007.

[196] A. Zayats, T. Kalkbrenner, V. Sandoghdar, and J. Mlynek. Second-harmonic generation from individual surface defects under local excitation. *Phys. Rev. B*, 61:4545, 2000.

[197] W. Zhang, H. Fischer, T. Schmid, R. Zenobi, and O.J.F. Martin. Mode-selective surface-enhanced Raman spectroscopy using nanofabricated plasmonic dipole antennas. *J. Phys. Chem. C*, 113:14672, 2009.

[198] J. Zheng, C. Zhang, and R.M. Dickson. Highly fluorescent,water-soluble, size-tunable gold quantum dots. *Phys. Rev. Lett.*, 93:077402, 2004.

[199] J.I. Ziegler and R.F. Haglund Jr. Plasmonic response of nanoscale spirals. *Nano Lett.*, 10:3013, 2010.

[200] P. Zijlstra, J.W.M. Chon, and M. Gu. Five-dimensional optical recording mediated by surface plasmons in gold nanorods. *Nature*, 459:410, 2009.

[201] P. Zijlstra, J.W.M. Chon, and M. Gu. White light scattering spectroscopy and electron microscopy of laser induced melting in single gold nanorods. *Phys. Chem. Chem. Phys.*, 11:5915, 2009.